2nd International Conference on Environmental Engineering and Sustainable Development (CEESD 2017)

IOP Conference Series: Earth and Environmental Science
Volume 112

Koh Samui, Thailand
8 – 10 December 2017

ISBN: 978-1-5108-5832-9
ISSN: 1755-1307

Printed from e-media with permission by:

Curran Associates, Inc.
57 Morehouse Lane
Red Hook, NY 12571

Some format issues inherent in the e-media version may also appear in this print version.

Printed by Curran Associates, Inc. (2018)

For permission requests, please contact the Institute of Physics
at the address below.

Institute of Physics
Dirac House, Temple Back
Bristol BS1 6BE UK

Phone: 44 1 17 929 7481
Fax: 44 1 17 920 0979

techtracking@iop.org

Additional copies of this publication are available from:

Curran Associates, Inc.
57 Morehouse Lane
Red Hook, NY 12571 USA
Phone: 845-758-0400
Fax: 845-758-2633
Email: curran@proceedings.com
Web: www.proceedings.com

Table of contents

Volume 112

2017 2nd International Conference on Environmental Engineering and Sustainable Development (CEESD 2017)

8–10 December 2017, Koh Samui, Thailand

Accepted papers received: 16 January 2018
Published online: 7 February 2018

Preface

Papers

Water quality, water and wastewater treatment technologies

PREFACE

2017 2nd International Conference on Environmental Engineering and Sustainable Development (CEESD 2017) were held in Koh Samui, Thailand, from December 8-10, 2017. The proceeding is a compilation of the accepted papers and represent an interesting outcome of the conference.

CEESD 2017 is organized by Asia Pacific Institute of Science and Engineering(APISE). The conferences provide a useful platform both for showing the latest research directions and for exchange of research results and ideas in broad Environmental Engineering and Sustainable Development area.

We would like to acknowledge all of those who supported CEESD 2017. Each individual and institutional help were very important for the success of this conference. Especially we would like to thank the organizing committee for their valuable advises in the organization and helpful peer review of the papers.

We hope that CEESD 2017 will be a forum for fruitful discussions that will shed light on new ideas and foster collaborative researches. We are sure that the proceedings will serve as an important research source of references and the knowledge, which will lead to not only scientific and engineering progress but also new products and processes.

Irene Man Chi Lo
The Hong Kong University of Science and Technology, HK

CEESD 2017 COMMITTEE

Conference Chair

Prof. Irene Man Chi Lo, The Hong Kong University of Science and Technology, HK

Program Chair

Prof. Weihua Zhang, Sun Yat-sen University, China

Prof. Liu Tongzhou, Harbin Institute of Technology Shenzhen Gradute School, China

Publication Chair

Assoc. Prof. Peng Wang, King Abdullah University of Science and Technology, Saudi Arabia

International Technical Committee

Prof. Rao Y. Surampalli, Global Institute for Energy, Environment and Sustainability, USA

Prof. Dr. Chih-Huang Weng, I-Shou University, Taiwan

Prof. Rao Pinhua, Shanghai Univeristy of Engineering Science, China

Prof. Wai Yuen SZETO, the University of Hong Kong, Hong Kong

Associate Prof. Zhang Zhen, South China Agricultural University, China

Dr. Sayed M. Bateni, University of Hawaii at Manoa, USA

Dr. Huichao Chen, Southeast University, China

Dr. Aniza Binti Abdul Aziz, Universiti Malaya, Malaysia

Dr. Huynh Khanh An, Ho Chi Minh City University of Technology, Vietnam

Peer review statement

All papers published in this volume of *IOP Conference Series: Earth and Environmental Science* have been peer reviewed through processes administered by the proceedings Editors. Reviews were conducted by expert referees to the professional and scientific standards expected of a proceedings journal published by IOP Publishing.

Relationship between Hydrodynamic Conditions and Water Quality in Landscape Water Body

Mengxin Kang[1,2,*], Yimei Tian[1], Haiya Zhang[1] and Dehong Wang[2]

[1] School of Environmental Science and Engineering, Tianjin University, Tianjin 300072, China
[2] College of Architecture Engineering, Northeast Electric Power University, Jilin 132012, China

*linxizimin@163.com

Abstract. The urban landscape water usually lacks necessary water cycle and water speed is closed to zero, which easily lead to eutrophication in water system and deterioration of water quality. Therefore, understanding the impact of water circulation on the water quality is of great significance. With that significance, this research has been done to investigate the relationship between hydrodynamic conditions and water quality of urban landscape water based on adopted water quality indexes such as chemical oxygen demand (COD), total nitrogen (TN), total phosphorus (TP) and nitrogen-ammonia (NH3-N). Moreover, MIKE 21 model is used to simulate the hydrodynamics and water quality under different cases in an urban landscape lake. The results of simulation show that water circulation system could effectively improve current speeds, reduce the proportion of stagnation area, and solve the problem of water quality deterioration caused by reclaimed water in the lake.

1. Introduction

Landscape water is indispensable to the city and serves to discharge flood waters, receive contaminants, provide scenery and support ecological systems. Most landscape waters are closed, with insufficient water mobility, weak hydrodynamic conditions, and fragile self-purification [1]. To relieve water shortages, unconventional water resources, such as reclaimed water, are used to recharge landscape water, but reclaimed water has high nitrogen and phosphorus concentrations, adding to the difficulty of guaranteeing the receiving water quality [2]. In order to keep and maintain landscape water quality, a hydrodynamic analysis is needed because: (1) good hydrodynamic condition increases the transport of oxygen from atmosphere to water and increases the concentration of dissolved oxygen; (2) the flow resuspends organic detritus deposited at the bottom, prolonging the organic detritus residence time, and enabling more complete aerobic degradation; (3) hydraulic structures or pumps that increase water flow enhance recreation and promote the overall ecological environment [3, 4].

In this paper, an urban landscape lake was selected to investigate the relationship between hydrodynamics and water quality, and MIKE 21 model was used to simulate the hydrodynamics and water quality under different cases.

2. Materials and Methods

2.1. Study Area

In this study, the artificial landscape Lake (Figure. 1) was used to simulate and predict water quality trends under different conditions. The lake has a surface area of 420,000 m^2, a maximum depth of 4 m, and a volume of 1,470,000 m^3. Reclaimed water will be used to recharge the lake and offset losses from evaporation and leakage (Table 1). Without external forcing, the wind-driven current is the only source of water mixing. COD, TN, TP and NH3-N are the primary water quality control targets.

Figure 1. Study area

Table 1. Water quality in the study area

water quality item	artificial lake	reclaimed water
NH3-N(mg/l)	0.53	8
TN(mg/l)	1.54	15
TP(mg/l)	0.08	0.5
COD(mg/l)	74	50

[a] No criteria.

2.2. Water Quality Sampling and Analysis

Water samples in the landscape Lake were collected daily from May 12 to June 24, 2014. The pH and DO were measured immediately in situ using a HACH HQ30d probe. The COD, TN, TP and NH3-N were analysed following the Water and Wastewater Monitoring and Analysis Method (China 2002). The atmospheric data were collected by the Tianjin Meteorological Bureau.

2.3. Model Description and Application

MIKE 21 is a two-dimensional mathematical model developed by DHI Water & Environment and could be used whenever stratification is neglected [5]. In this paper, MIKE 21 was used to simulate hydrodynamics and water quality, and to forecast water quality in the landscape Lake.

The water environment model is composed of the hydrodynamic model and advection/dispersion models [6]. The control equations of the hydrodynamic model are the 2-D shallow water equations in Cartesian coordinates:

$$\frac{\partial h}{\partial t}+\frac{\partial h\bar{u}}{\partial x}+\frac{\partial h\bar{v}}{\partial y}=hS \tag{1}$$

$$\frac{\partial h\bar{u}}{\partial t}+\frac{\partial h\bar{u}^2}{\partial x}+\frac{\partial h\overline{vu}}{\partial y}=f\bar{v}h-gh\frac{\partial \eta}{\partial x}-\frac{h}{\rho_0}\frac{\partial p_a}{\partial x}-\frac{gh^2}{2\rho_0}\frac{\partial \rho}{\partial x}+\frac{\tau_{sx}}{\rho_0}-\frac{\tau_{bx}}{\rho_0}-\frac{1}{\rho}\left(\frac{\partial s_{xx}}{\partial x}+\frac{\partial s_{xy}}{\partial x}\right)+\frac{\partial}{\partial x}\left(hT_{xx}\right)+\frac{\partial}{\partial x}\left(hT_{xy}\right)+hu_sS \tag{2}$$

$$\frac{\partial h\bar{v}}{\partial t}+\frac{\partial h\overline{uv}}{\partial x}+\frac{\partial h\bar{v}^2}{\partial y}=-f\bar{u}h-gh\frac{\partial \eta}{\partial y}-\frac{h}{\rho_0}\frac{\partial p_a}{\partial y}-\frac{gh^2}{2\rho_0}\frac{\partial \rho}{\partial y}+\frac{\tau_{sy}}{\rho_0}-\frac{\tau_{by}}{\rho_0}-\frac{1}{\rho_0}\left(\frac{\partial s_{yx}}{\partial y}+\frac{\partial s_{yy}}{\partial x}\right)+\frac{\partial}{\partial x}\left(hT_{xy}\right)+\frac{\partial}{\partial y}\left(hT_{yy}\right)+hv_sS \quad (3)$$

Equation (1) is the continuity equation, and equations (2) and (3) are the momentum equations in the x and y directions, respectively.

The water quality model uses the advection-dispersion equation (4) and assuming the degradation processes satisfy the first order reaction mechanism

$$\frac{\partial C}{\partial t}+u\frac{\partial C}{\partial x_i}+v\frac{\partial C}{\partial y_i}=D_x\frac{\partial^2 C}{\partial x^2}+D_y\frac{\partial^2 C}{\partial y^2} \quad (4)$$

$$\frac{\partial C}{\partial t}=-KC \quad (5)$$

3. Results and Discussion

The study area was sensitive to the environmental conditions, there was no natural inflows or outflows, and rainfall cannot replace water lost by evaporation and leakage. Therefore, reclaimed water would be added to maintain the water level. The reclaimed water will flow into the lake from two points: z1, in the northern part of the lake, and z2 in the southern part of the lake (Figure. 1). Compared with the lake water, the reclaimed water had higher concentrations of nitrogen and phosphorus (Table 1). The reclaimed water was equivalent to point source pollution, and easily cause accumulations of nitrogen and phosphorus and subsequent eutrophication. Two canals (north canal and south canal) were planned to build along the lake, and one part of the lake water would flow from point x1, through the north canal and return to the lake at points x2, x3, x4, x5, x6, and the other part of the water would flow from source x10, through the south canal and return to the lake at points x7, x8, x9 (Figure 1). The lake and the water canals would form water circulation system to improve the water hydrodynamic condition. The velocity of the inlet point x2, x3, x4, x5, x6, x7, x8, x9 was 0.0288 m/s, 0.0817 m/s, 0.2655 m/s, 0.1745 m/s, 0.2120 m/s, 0.0648 m/s, 0.1763 m/s, 0.0827 m/s, respectively. And the recirculation system flow rate was $0.81 m^3/s$.

3.1. Model Calibration

Measurements of water quality and velocity from May 12, 2015 to June 30, 2015 and relevant meteorological parameters were used to calibrate the model. The simulation agreed with the observed values within 10%. These results indicated that the MIKE 21 model was properly calibrated and could be used to study the principal water quality characteristics.

3.2. Water Quality Simulation

In this study, the artificial landscape Lake (Figure.1) was used to simulate and predict water quality trends under different conditions. The flow field distribution in the lake was shown in Figure 2. The average current speed was 0.0034 m/s if only by wind (Figure.2a), while the average current speed was 0.008 m/s if the water circulation system built (Figure.2b).

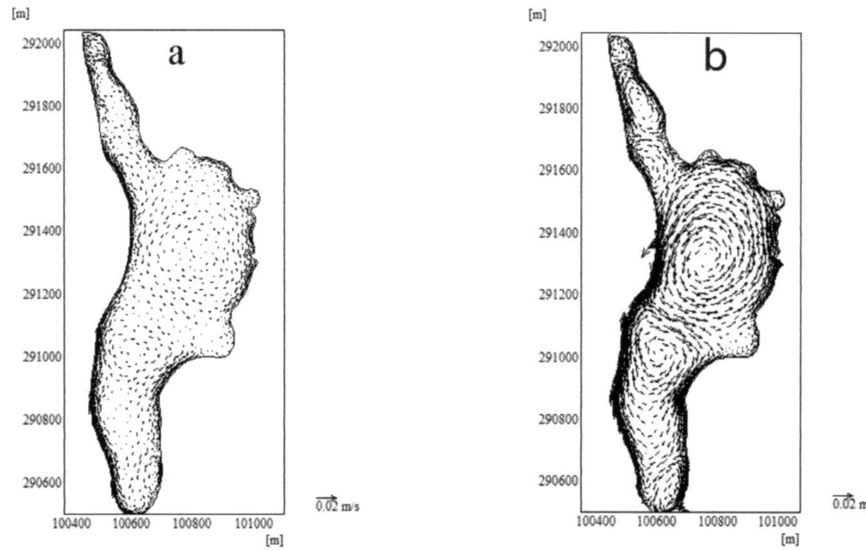

Figure 2. Flow field distribution in the lake

In this article, the influences of circulation system and reclaimed water recharging system on the lake water quality were simulated. Case 1 had only recharging system but no circulation system; and case 2 had both circulation and recharging systems. The two cases simulated the water quality for July 2016. The calculated concentrations of COD, TN, TP and NH3-N in the north, mid, south of the lake were shown in Figure. 3a-h.

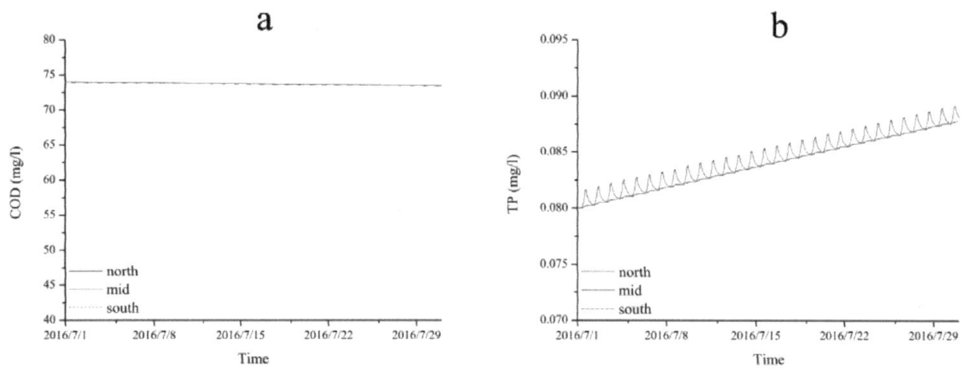

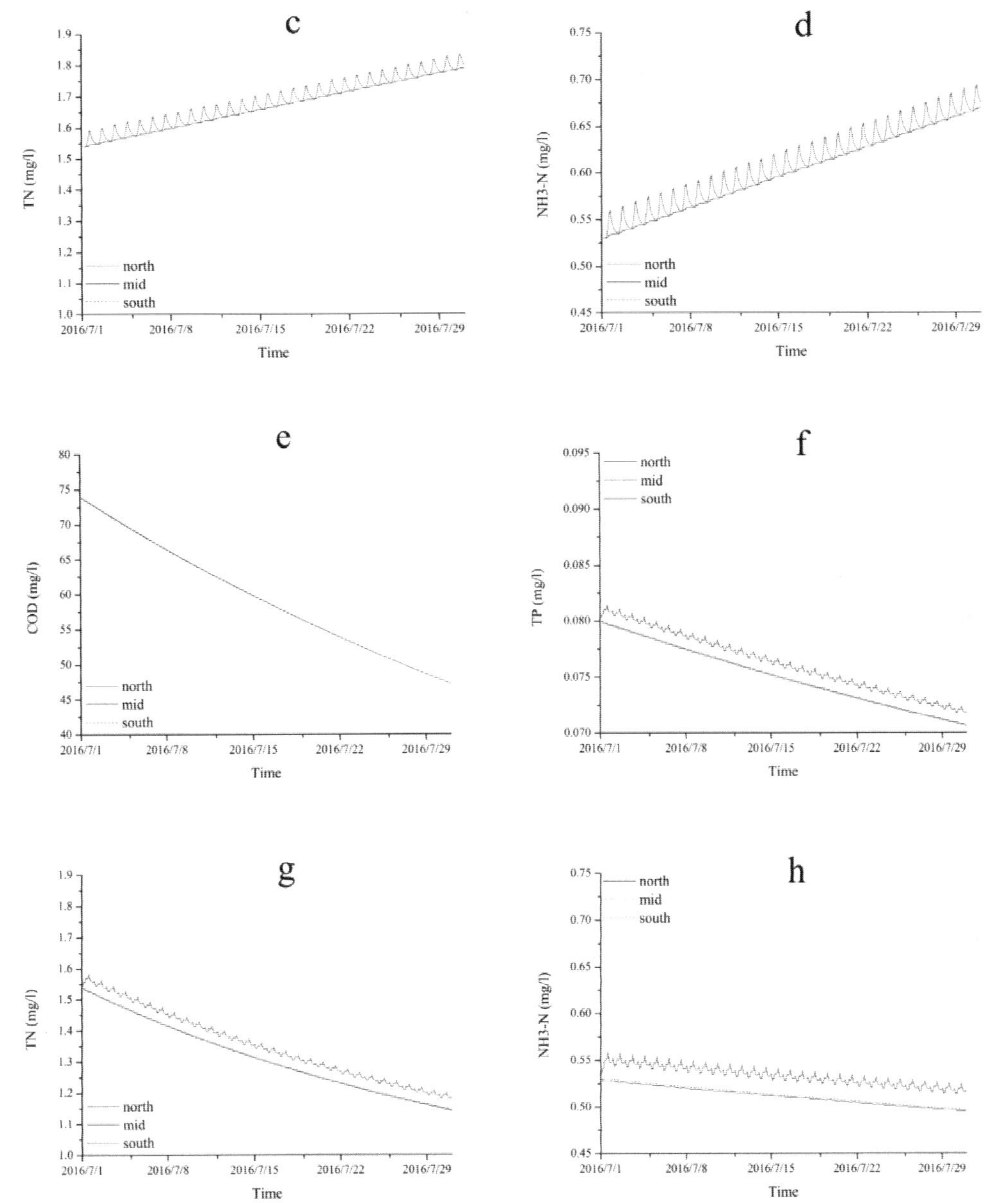

Figure 3. Simulation results for the two cases

In Case 1, the reclaimed water was added to the lake from the points z1 and z2, about 29500 m^3/d. Figure. 3a-d showed the water quality for Case 1. The reclaimed water and the lake water COD concentrations was not significantly different, and the relative volume of reclaimed water was small (approximately 3%), so the change of COD was small. The concentrations of TN, NH3-N, and TP in the reclaimed water were approximately 15, 6 and 10 times larger than in lake, respectively, therefore, the changes in those concentrations were readily observed. The calculated concentrations at the north part were strongly affected by the addition of reclaimed water, because the north part of the lake was narrow with poor dispersion. The south and mid part ware rather open and had strong convective diffusion, so the pollutant concentrations were lower than the north part. When the simulation completed, the concentrations of COD, TN, NH3-N, TP were 73.5 mg/l, 1.8mg/l, 0.68mg/l and 0.088mg/l, respectively. These results show that if the lake was recharged with reclaimed water (without a circulation system), nitrogen and phosphorus would accumulate and led to eutrophication.

Case 2 included both the circulation and recharging systems. The circulation system, which was favourable for the dilution and diffusion of pollutant, rapidly mixed the recharge water throughout the lake. And through the continuous renew of water-air interface, the DO content was increased and the self-purification ability was strengthened. The water quality was consistent at all three parts (Figure. 3e-h). Variation at the north part was smaller than in Case 1. After the simulation completed, the concentrations of COD, TN, NH3-N, TP were 47 mg/l, 0.115 mg/l, 0.5 mg/l and 0.07 mg/l, respectively. Compared with Case1, the concentrations had decreased by 37%, 35%, 27% and 20%, respectively.

4. Conclusions

This paper presented the relationship between hydrodynamics and water quality of urban landscape water. MIKE 21 model was used to simulate the hydrodynamics and water quality under different cases in an urban landscape lake. The simulation result showed that a water circulation system could effectively improve current speed and solve the problem of water quality deterioration caused by reclaimed water added to offset water losses from evaporation and leakage in the lake. And based on the mechanism of relationship of hydrodynamic and water quality, the Olympic forest park in Beijing and the living water garden in Chengdu have already achieved the expected outcomes. Therefore, a circulation system could effectively maintain water quality and eliminate the risk of water eutrophication when reclaimed water was used to recharge the lake.

Acknowledgments
The work was supported by the "Major Science and Technology Program for Water Pollution Control and Treatment" (2014ZX07203-009).

References

[1] Dupas R, Delmas M, dorioz J-M, et al. 2015 Assessing the impact of agricultural pressures on N and P loads and eutrophication risk. Ecological Indicators, 48(p396-407).
[2] Yang L, He J, Liu Y, et al. 2016 Characteristics of change in water quality along reclaimed water intake area of the Chaobai River in Beijing, China. Journal of environmental sciences, 50(p93-102).
[3] Huang L, Fang H, He G, et al. 2016 Effects of internal loading on phosphorus distribution in the Taihu Lake driven by wind waves and lake currents. Environmental pollution, 219(p760-73).
[4] Zeng Q, Qin L, Li X. 2015 The potential impact of an inter-basin water transfer project on nutrients (nitrogen and phosphorous) and chlorophyll a of the receiving water system. The Science of the total environment, 536(p675-86).
[5] Li Y, Zhang Q, Yao J. 2015 Investigation of Residence and Travel Times in a Large Floodplain Lake with Complex Lake-River Interactions: Poyang Lake (China). Water, 7(5): p1991-2012.
[6] Li-kun Y, Sen P, Xin-hua Z, et al. 2017Development of a two-dimensional eutrophication model in an urban lake (China) and the application of uncertainty analysis. Ecological Modelling, 345(p63-74).

A Surfactant-Free Synthesis Technique of Coral-Like Zno Hierarchical Structures for Photocatalytic Degradation of Resorcinol under UV Irradiation

Jin-Chung Sin[1,*], Sze-Mun Lam[2] and Ying-Hui Chin[1]

[1] Department of Petrochemical Engineering, Faculty of Engineering and Green Technology, Universiti Tunku Abdul Rahman, Jalan Universiti, Bandar Barat, 31900 Kampar, Perak, Malaysia

[2] Department of Environmental Engineering, Faculty of Engineering and Green Technology, Universiti Tunku Abdul Rahman, Jalan Universiti, Bandar Barat, 31900 Kampar, Perak, Malaysia

*sinjc@utar.edu.my

Abstract. Hierarchical coral-like ZnO structures were successfully prepared by a surfactant-free wet chemical method. Various characterization tests were carried out to analyze the as-prepared ZnO samples. The coral-like ZnO was used to degrade resorcinol at three different solution pH values (pH 5.0, pH 8.0 and pH 11.0). It was observed that the resorcinol adsorption onto the ZnO was strongly dependent on the electrical charge properties of both photocatalyst and resorcinol. Photocatalytic degradation of resorcinol reached the highest at pH 11.0 due to high concentration of hydroxyl ions for hydroxyl radicals generation.

1. Introduction

Advanced oxidation process using semiconductor photocatalysts has garnered much interest in recent years due to its efficiency of conversion of solar energy into chemical energy in various applications including degradation and complete elimination of environmental pollutants [1,2]. Among various oxide semiconductor photocatalysts, considerable efforts have been made to synthesize diverse nanoscale semiconductor materials (nanorods, nanoparticles, nanoneedles and nanosheets) with controlled morphologies and dimensions [3,4]. Recently, hierarchical structures combined the features of nanoscale building blocks have been found to be attractive class of materials owing to their unique properties and peculiar structure different from those of mono-morphological structures [3,5,6]. Studies have revealed that hierarchical structures were more structurally stable to hinder the aggregation of particles and offered larger number of surface activation sites for photocatalytic reaction [3,5]. Many hierarchical structures such as flower-shaped, urchin-shaped, spherical-shaped and dandelion-shaped have been synthesized via different methods [5-9]. However, most of the hierarchical structures were prepared with the aid of surfactant or template, which were difficult to obtain low-cost mass production.

ZnO is a II-IV compound semiconductor with direct wide band gap of ~3.3 eV, which has been widely used as significant photocatalytic material due to its photochemical and catalytic properties together with its low cost [10]. The present work thus focuses on the surfactant-free wet chemical method to synthesize ZnO hierarchical structures with coral-like morphology for photocatalytic degradation of resorcinol aqueous solution. Resorcinol was chosen since it is a well-known endocrine disrupting chemical and extensively used in the dyes, adhesives and pharmaceuticals manufacturing. The resorcinol effect on rats and humans have found the disruptions of thyroid activity via the interference of triiodothyronine (T3) and thyroxine (T4) metabolism [11].

2. Experimental

All the chemical reagents were analytical-grade purity and used as purchased without further purification. The details for the synthesis of hierarchical coral-like ZnO structures were as follows: zinc nitrate hexahydrate and sodium hydrate with the molar ratios 1:6 were dissolved in 80 mL deionized water. After 3 h stirring, the resulting white precipitates were collected by filtration, washed with deionized water and followed by drying in air at 60°C for 12 h. The dried precipitates were finally calcined at 650°C for 2 h.

The obtained samples were characterized by field-emission scanning electron microscopy (FESEM, Quanta FEG 450), energy dispersion X-ray (EDX, Oxford X-max), X-ray diffraction (XRD, Philips PW1820 diffractometer), UV-vis diffuse reflectance spectroscopy (UV-vis DRS, Perkin Elmer Lambda 35) and nitrogen adsorption-desorption (Micromeritics ASAP 2020).

The photocatalytic experiment was conducted in an immersion well photoreactor (height: 200 mm, outer diameter: 100 mm, inner diameter: 60 mm). A 15 W UV Pen-Ray (UVP) lamp (λ_{max} = 365 nm) as UV source was placed in the center of cylindrical photoreactor. The photon intensity was measured as 0.840 $mWcm^{-2}$ at distance 10 mm away from the UV lamp. The experiments were conducted with the initial resorcinol concentration of 20 mg/L and 350 mg of ZnO catalysts. The ZnO catalysts were added to the photoreactor containing 350 mL of resorcinol solution. The pH adjustment was done using equimolar NaOH and HCl solutions. Before the photocatalytic reaction, the mixture was equilibrated under darkness for 60 min to reach adsorption-desorption equilibrium. Then, the suspension was exposed to UV irradiation. 5 mL suspensions were collected after every 30 min irradiation and then centrifuged to remove the solid particles. The concentrations of resorcinol were monitored using a high performance liquid chromatography (HPLC, Perkin Elmer Series 200). The HPLC separation was carried out using a mobile phase composed of acetonitrile-water mixtures at wavelength of 238 nm. In order to determine the reproducibility of all the results, at least duplicated runs were carried out for each condition for averaging the results, and the experimental error was found to be within ±4%.

3. Results and discussion

The as-prepared ZnO samples were morphologically characterized by the FESEM. The highly dense coral-shaped structures were observed in Figure 1(a). The magnified image in Figure 1(b) clearly displayed that the structure of the coral-like ZnO was accumulated by lot of nanosheets and the thickness of the nanosheets ranged from 50-132 nm. In addition, the nanosheets interlaced with each other and formed porous hierarchical coral-like ZnO structures through oriented aggregation. Figure 1(c) depicts the EDX spectrum of as-prepared ZnO samples. The Zn and O peaks were observed, revealing the as-prepared samples were pure ZnO. A signal of C was attributed to the adhesive carbon tape used in the EDX analysis. In support, XRD analysis of the as-prepared ZnO samples was also conducted to investigate the crystalline phases as shown in Figure 1(d). The diffraction peaks in the XRD pattern can be indexed to a hexagonal wurtzite structure. The sharp diffraction peaks and no impurity phases were detected in the XRD pattern, which confirmed the well-crystallinity and purity of the as-prepared ZnO samples.

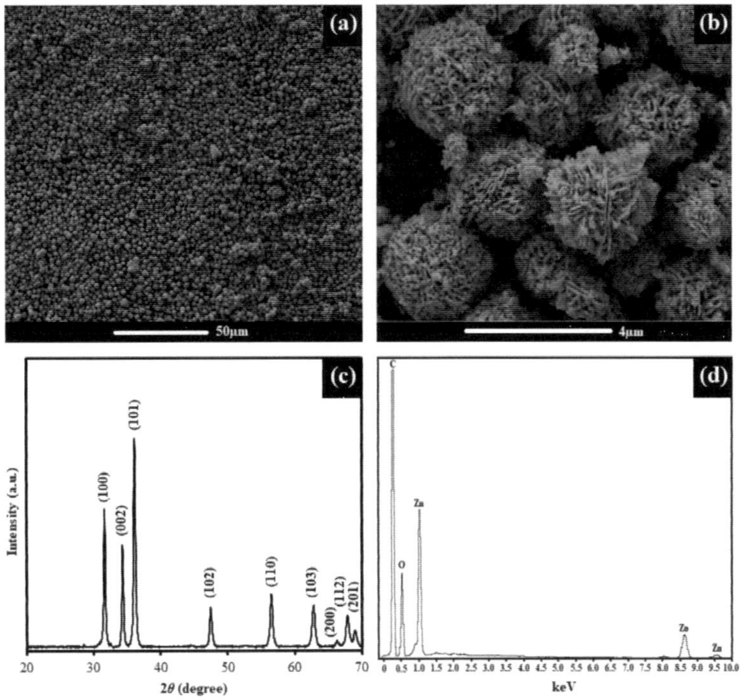

Figure 1. (a) Low-magnification and (b) high-magnification FESEM images, (c) EDX spectrum and (d) XRD pattern of as-prepared hierarchical coral-like ZnO structures.

UV-vis DRS analysis was conducted to elucidate the optical properties of as-prepared samples. From Figure 2(a), a strong absorption in the UV region which lay between 370 and 390 nm was observed. This was a characteristic and significant region of the absorption band of the hexagonal wurtzite structure of ZnO [12]. The band gap energy was determined using E_g (eV) = 1240/λ (nm), where E_g is the band gap energy and λ is the wavelength (nm) of absorption onset. The E_g of the samples was measured to be 3.22 eV, which was close to the values reported in literatures [4,12]. Full nitrogen sorption isotherms were also determined to obtain the specific surface area and the pore size of the as-prepared ZnO samples as shown in Figure 2(b). The obtained isotherms were identified as type IV, showing the existence of abundant mesopores in the ZnO samples. From the corresponding pore size distribution curve in the inset of Figure 2(b), the mesoporous with peaks at maximum points of 2 nm and 29 nm were indicated. The specific surface area was evaluated to be 6.5 m^2/g from data point by the Brunauer–Emmett–Teller (BET) method.

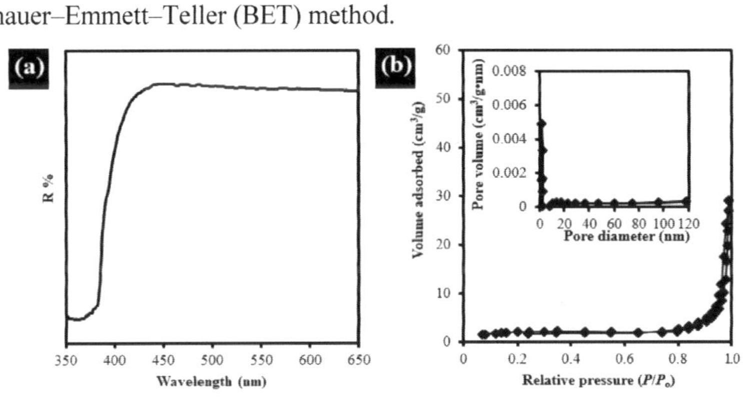

Figure 2. (a) UV-vis DRS spectrum and (b) N_2 adsorption–desorption isotherms of the as-prepared hierarchical coral-like ZnO structures. Inset of (b) is the pore size distribution.

Photogeneration of electron-hole pairs is generally played pivotal role in the photocatalytic degradation of organic pollutants. The photogenerated hole (h^+) in the valence band can trap on the ZnO surface and reacted with either surface bound hydroxide species or adsorbed water molecules through the following reactions [13]:

$$h^+ + OH^- \rightarrow \bullet OH \quad (1)$$

$$h^+ + H_2O \rightarrow \bullet OH + H^+ \quad (2)$$

At the same time, the photogenerated electron (e^-) in the conduction band of ZnO reacted with adsorbed oxygen on the ZnO surface to promote the superoxide anion radical ($O_2\bullet^-$) formation and then converted to active hydroxyl radical (•OH). The •OH, $O_2\bullet^-$ and h^+ formed by ZnO via e^-- h^+ generation under light irradiation, which are highly active species to break the organic pollutants into CO_2 and H_2O [13,14].

$$e^- + O_2 \rightarrow O_2\bullet^- \quad (3)$$

$$O_2\bullet^- + H^+ \rightarrow \bullet OOH \quad (4)$$

$$\bullet OOH + H^+ + e^- \rightarrow H_2O_2 \quad (5)$$

$$H_2O_2 + e^- \rightarrow \bullet OH + OH^- \quad (6)$$

$$\text{Organic Pollutants} + (\bullet OH,\ O_2\bullet^- \text{ and } h^+) \rightarrow \text{degradation intermediates} \rightarrow CO_2 \text{ and } H_2O \quad (7)$$

Figure 3(a) shows the resorcinol degradation over hierarchical coral-like ZnO structures schematically. Figure 3(b) depicts the adsorption of resorcinol onto the surface of as-prepared ZnO samples at different solution pH values. It was observed that increase in solution pH from 5.0 to 11.0 decreased the adsorption efficiency of resorcinol. Since the zero potential point of ZnO was 9.3 [11], the ZnO surface was positively charged at pH 5.0 and an electrostatic attraction between positively charged ZnO with molecular form of resorcinol molecule led to maximum resorcinol adsorption efficiency. In contrast, the ZnO surface deprotonated at higher pH and formed repulsive forces with the negatively charged resorcinol. The degradation process mainly happened at the interface between the catalyst surfaces and organic pollutants, thus the adsorption of resorcinol is an important step in photocatalysis. Even though the organic pollutants with high adsorption efficiency degraded faster, the effect sites for UV light absorption decreased with increasing the adsorption. It should also be noted that higher pH value can provide large amount of OH^- ions on the ZnO surface and favoured the formation of •OH, which then led to photocatalytic enhancement of resorcinol. Figure 3(c) shows the degradation percentage of resorcinol at each irradiation time for different solution pH values. It can be observed that the concentration of resorcinol gradually decreased with increasing the irradiation time at all investigated pH values. The kinetic studies of resorcinol degradation over the as-prepared ZnO samples are shown in Figure 3(d). The photocatalytic reactions of resorcinol were in agreement with the first-order kinetic behaviour according to the Langmuir-Hinshelwood model and can be stated as [11]:

$$\ln(C_o/C) = kt \quad (8)$$

where k is the observed rate constant, C_o is the equilibrium concentration of resorcinol and C is the concentration at time t. The k values for the photocatalytic reactions at pH 5.0, pH 8.0 and pH 11.0 were measured to be 0.0074, 0.0136 and 0.0271 min^{-1}, respectively. The k value of photocatalytic reactions over coral-like ZnO at pH 11.0 was 3.7-fold and 2.0-fold higher than those of photocatalytic reactions at pH 5.0 and pH 8.0, respectively. Therefore, even though the adsorption of resorcinol was low at alkaline pH, the explanation for this behaviour was the presence of higher concentration of OH^- ions or higher fraction of hydrolyzed forms of resorcinol, which can lead to higher concentration of •OH radicals to enhance the photodegradation rate.

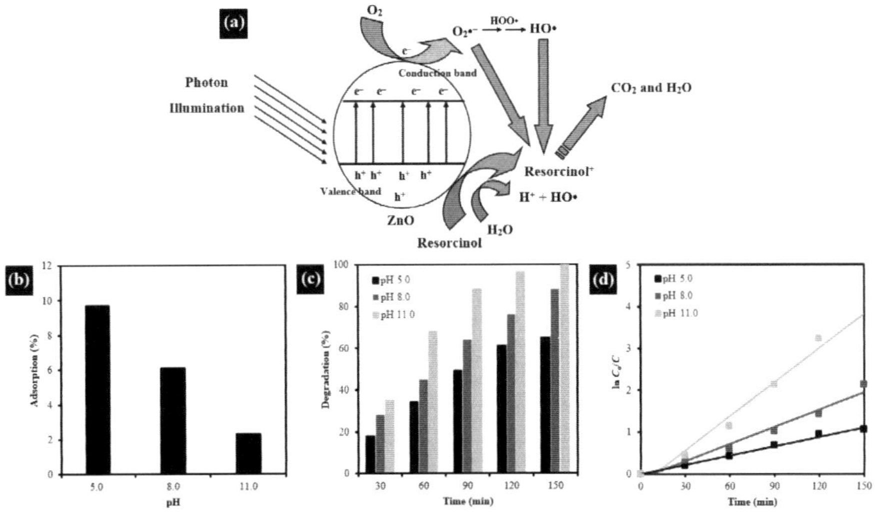

Figure 3. (a) Schematic illustration of resorcinol degradation over hierarchical coral-like ZnO structures, (b) Adsorption of resorcinol onto ZnO at different pH values, (c) Degradation percentage of resorcinol as a function of irradiation time and (d) First-order linear plots of $\ln(C_o/C)$ versus irradiation time.

4. Conclusions

Hierarchical coral-like ZnO structures were successfully synthesized via a surfactant-free wet chemical method. The as-prepared ZnO samples were characterized using FESEM, EDX, XRD, UV-vis DRS and nitrogen adsorption-desorption measurements. The photocatalytic degradation of resorcinol over as-prepared ZnO samples was carried out at three solution pH values (pH 5.0, pH 8.0 and pH 11.0). The results showed that the extent of resorcinol adsorption onto the ZnO strongly dependent on the electrical charge properties of both photocatalyst and resorcinol. High photodegradation of resorcinol was obtained at pH 11.0 with the high concentration of OH^- ions for •OH radicals generation.

Acknowledgments

This work was supported by Universiti Tunku Abdul Rahman (UTARRF/2016-C2/S03 and UTARRF/2017-C1/L02) and Ministry of Higher Education of Malaysia (FRGS /1/2015/TK02/UTAR /02/2 and FRGS/1/2016/TK02/UTAR/02/1).

References

[1] Zhao Y, Tan X, Yu T and Wang SC 2016 *Mater. Lett.* **164** 243
[2] Bao YP, Lim TT, Zhong ZY, Wang R and Hu X 2017 *J. Colloid Interface Sci.* **505** 489
[3] Lu F, Cai WP and Zhang YG 2008 *Adv. Funct. Mater.* **18** 1047
[4] Lam SM, Sin JC and Mohamed AR 2016 *Mater. Lett.* **167** 141
[5] Yu CY, Yu YL, Xu T, Wang XL, Ahmad M and Sun HY 2017 *Mater. Lett.* **190** 185
[6] Sin JC, Lim CA, Lam SM and Mohamed AR 2017 *J. Mater. Sci.: Mater. Electron.* **28** 13236
[7] Taheri M, Abdizadeh H and Golobostanfard MR 2017 *J. Alloys Compd.* **725** 291
[8] Zhang FN, Zhao Y, Li YW, Wu GJ and Zhao JZ 2017 *J. Colloid Interface Sci.* **505** 519
[9] Yue XZ, Yi SS, Wang RW, Zhang ZT and Qiu SL 2017 *J. Mater Chem. A* **5** 10591
[10] Lam SM, Quek JA and Sin JC 2017 *Mater. Lett.* **195** 34
[11] Sin JC, Lam SM, Lee KT and Mohamed AR 2013 *Desalin. Water Treat.* **51** 3505
[12] Lam SM, Sin JC, Abdullah AZ and Mohamed AR 2013 *Mater. Lett.* **93** 423
[13] Reza W, Faisal SM, Owais M, Bahnemann D and Muneer M 2016 *RSC Adv.* **6** 78335
[14] Shakir M, Faraz M, Sherwani MA and Al-Resayes SI 2016 *J. Lumin.* **176** 159

Fabrication of Flower-like ZnO Micro/Nanostructures for Photodegradation of Pre-treated Palm Oil Mill Effluent

Sze-Mun Lam[1*], Kok-Ann Wong[1] and Jin-Chung Sin[2]

[1] Department of Environmental Engineering, Faculty of Engineering and Green Technology, Universiti Tunku Abdul Rahman, Jalan Universiti, Bandar Barat, 31900 Kampar, Perak, Malaysia

[2] Department of Petrochemical Engineering, Faculty of Engineering and Green Technology, Universiti Tunku Abdul Rahman, Jalan Universiti, Bandar Barat, 31900 Kampar, Perak, Malaysia

*Email: lamsm@utar.edu.my

Abstract. Flower-like ZnO micro/nanostructures were fabricated by a simple and surfactant-free reflux method. X-ray diffraction findings showed that the prepared ZnO product was highly crystallite with hexagonal wurtzite structure. The band gap energy of ZnO sample was measured to be 3.18 eV via an optical reflectance spectrum. The flower-like morphological features of ZnO micro/nanostructures were witnessed through field-emission scanning electron microscopy. Such micro/nanoparticles could be used in the photodegradation of pre-treated palm oil mil effluent (POME) under UV irradiation.

1. Introduction

To date, search suitable and sustainable pollution-degradation strategies become paramount important to circumvent the environmental problems. Particularly, Malaysia as one of the largest water polluters originates from oil palm industries. It is a well-known that fresh POME is acidic with the pH ranging from 4 to 5, chemical oxygen demand (COD) of 40,000–100,000 mg/L and biological oxygen demand (BOD) of 25,000–65,000 mg/L [1–3]. Direct release of palm oil mill effluent (POME) causes severe environmental pollution [1,2]. The traditional POME treatment in Malaysia is through aerobic ponding system. However, there were some shortcomings of this treating method including land consuming and long retention time (up to months).

An alternative approach, light irradiation/semiconductor technology has been recently demonstrated for POME degradation [3–5]. This technique employs light energy as an activating agent for photocatalyst light absorption and initiated a series of active radicals generation for photo-mineralization of organic compounds. ZnO is a semiconductor which is extensively studied for a myriad of applications such as hydrogen-evolution, air pollution control and water treatment and others, attributed to its unique electrical, optical and mechanical properties [5,6]. Due to its saline features, a myriad of wet chemical technique including co-precipitation, sol-gel, hydrothermal techniques and some wet-chemical solution processes have been widely utilized to fabricate various ZnO nanostructures. It is well known that the nanoscale photocatalyst is anticipated to improve photocatalytic performance because of a larger specific surface area and more surface contact sites for the process [7]. Among these available techniques, reflux method is a facile, cheap, effective method and most importantly it can be conducted at short periods, normal pressure and at room temperature. Despite this method has been used to prepare the ZnO nanoparticles, its ability to form different structures at adequately huge amounts for practical applications including photocatalysis still deserves to be investigated.

In the present study, flower-like ZnO micro/nanoparticles has been fabricated by a reflux technique at low temperature (~65°C). The ZnO micro/nanoparticles were then characterized by various characterization techniques such as X-ray diffraction, field emission-scanning electron microscopy, energy dispersive X-ray spectroscopy, Fourier transform infrared spectroscopy and UV–vis diffuse reflectance spectroscopy measurements. The photocatalytic activity of ZnO micro/nanoparticles were evaluated in pre-treated POME under UV irradiation.

2. Experimental

All chemicals were in analytical grade and used as-purchased without further purification. Distilled water (DI) was utilized during the study. Pre-treated POME was collected from a palm oil mill factory positioned in Air Kuning, Perak Malaysia and stored in 4°C freezer.

2.1. Preparation of ZnO micro/nanoflowers

$Zn(NO_3)_2{\cdot}6H_2O$ (0.1 M) was first prepared in 100 mL distilled water and stirring continuously for 20 min. The mixture was then slowly added into NaOH (0.1 M) solution under magnetic stirring. The final pH value of solution was adjusted to 12.12 by addition of extra NaOH. Subsequently, the mixed suspensions were heated and refluxed 60°C for 8 h and cooled to room temperature naturally. The as-formed products were filtered, washed with distilled water and ethanol and dried at room temperature.

2.2. Characterization

The samples were analyzed by X-ray diffraction (XRD) pattern recorded using a Philips PW1820 diffractometer with Cu Kα radiation at a scanning rate of 2° min^{-1} in the range of 20°–80°. The morphology of the samples was scanned using a Quanta FEG 450 field emission-scanning electron microscope (FESEM) with an energy dispersive X-ray spectroscopy (EDX) analysis. UV–vis diffuse reflectance spectroscopy (DRS) spectrum was recorded on a Perkin Elmer Lambda 35 UV-vis spectrophotometer with $BaSO_4$ as a reference. The infrared spectrum was tested using a Thermo scientific Nicolet IS10 Fourier Transform Infrared (FTIR) spectrometer at scanning range from 500 to 4000 cm^{-1}. Photoluminescence (PL) spectroscopy was conducted at room temperature using a Perkin Elmer Lambda S55 spectrofluorometer using a Xe lamp with an excitation wavelength of 315 nm.

2.3. Photocatalytic test

All experiments were carried out in an immersion well photoreactor with dimensions 200 × 100 × 60 mm (height × outer diameter × inner diameter). In the center of the photoreactor, a 15 W UV Pen-Ray lamp with maximum emission at approximately 254 nm was used as UV source. The total UV output intensity at distance of 10 mm from the UV light, measured by use of a radiometer was 5400 $\mu W/cm^2$. The reaction mixture was prepared by adding 400 mg catalyst to the photoreactor containing 400 mL of pre-POME solution. Before the photoreaction, the pre-POME solution was magnetically stirred in the dark for 1 h. It was then irradiated and bubbled with air at a fixed flow rate of 3 mL/min. At different time intervals, 5 mL of the solution was withdrawn and centrifuged immediately to separate the suspended solids. The filtrate was then analyzed by COD, BOD_5, total suspended solid (TSS), oil and grease, turbidity, colour and pH in according to the standard methods of the American Public Health Association (APHA) [8]. The COD was determined using a Hach DRB200COD reactor after 2 h of digestion at 423 K. To analyze BOD_5, the samples were subjected to an incubator for 5 days at 298 K. Prior to the incubation, 10 mL of samples were diluted to 300 mL with deionized water. The dissolved oxygen (DO) for BOD_5 was taken before and after the incubation process. The BOD_5 value can be calculated as shown in Equation (1):

$$BOD_5\ (mg/L) = (DO_1 - DO_2)/P \tag{1}$$

where DO_1 and DO_2 are the initial and final DO values, respectively and P denotes as the dilution factor. In addition, the turbidity was measured using a Lovibond Turbicheck turbidimeter and expressed in Nephelometric Turbidity Unit (NTU). The TSS was determined according to 8006 Photometric method [9] using an UV-Vis spectrophotometer. The TSS content was expressed in mg/L. The colour of the sample was also determined using an UV-Vis spectrophotometer according to a colorimetric platinum cobalt method as measured in platinum cobalt (Pt-Co) unit. To study the oil and grease content in the pre-treated POME, three doses of hexane extraction was performed on 30 mL of pre-treated POME sample and evaporated in water bath at 60°C. The weight of round bottom flask before and after evaporated was recorded and the difference between the weights was the oil and grease content in pre-treated POME sample. The solution pH of the sample was tested using a HANNA Instruments HI2550 pH meter. In order to determine the reproducibility of the results, at least duplicated runs were carried out for each condition for averaging the results and the experimental error was found to be within ±4%.

3. Results and discussion

Figure 1a displays the XRD pattern of the as-prepared ZnO product. All of the diffraction peaks were labelled and could be matched with hexagonal wurtzite phase of ZnO (JCPDS Card file No. 36–1451) [10,11]. The strong and sharp XRD peaks in the figure demonstrated that the as-prepared ZnO were highly crystallinity. Further confirmation of formation of ZnO sample obtained from the EDX analysis. The presence of Zn and O peaks can be simply found in Figure 1b. The elemental carbon (C) detected was came from the supporting carbon tape. The UV–vis DRS spectrum of the ZnO product (Figure 1c) depicted a steep absorption edge which lay around 400 nm without any other absorption peak. The obtained region was a characteristic region for the absorption band of the hexagonal phase of ZnO [9,11]. The band gap energy of the ZnO sample can be acquired from the plot of $(F(R)hv)^{0.5}$ versus photon energy (E). The band gap energy was measured to be 3.18 eV (inset of figure 1c), which was concordance with the reported ZnO catalyst [12].

Figure 1d presents FTIR spectrum of as-prepared ZnO products. The emergence of distinctive absorption band around ~500 cm^{-1} could be related to Zn–O stretching vibrations. A distinct band appearing around 898 cm^{-1} in the samples can be assigned to the nitrate (NO_3^-) ion groups. Moreover, a weak band at 1620 cm^{-1} could be corresponded to bending modes of O–H vibration, whereas an intense broad band at around 3410 cm^{-1} for the samples can be attributed to O–H stretching vibration [13,14]. These FTIR results confirmed that ZnO micro/nanoflowers possessed hexagonal structures, which was in agreement with the results of XRD (figure 1a).

FESEM analysis of as-prepared ZnO products was carried out under different magnifications. Figure 2a shows the production of flower-like ZnO micro/nanostructures with high yield. The sizes of the products were quite uniform and varied from 800 nm to 2.6 µm. The magnified image in figure 2b indicated that flower-like micro/nanostructure with several petals growing radially from the center. Thus, a simple, cost-effective and surfactant-free chemical solution technique was successfully used to produce a large amount synthesis of 3D ZnO micro/nanostructures in this work.

The photoactivity of as-prepared ZnO samples was examined by degrading pre-treated POME under UV light irradiation. Figure 3 shows that the POME solution was degraded about 96.0% by ZnO micro/nanoflowers under 120 min irradiation time. Further experiments were conducted to examine the photoactivity of photocatalysts. When the solution with the micro/nanoflowers was kept in the dark or when the solution was irradiated without the ZnO micro/nanoflowers, only small amounts of pre-treated POME were adsorbed or degraded. The degree of pre-treated POME degradation of the two experiments attained 2.1% and 18.8%, respectively under 150 min irradiation time. Under similar conditions of UV light irradiation, the ZnO micro/nanoflowers indicated much greater activities that of commercial ZnO (63.0%). It inferred that the as-prepared ZnO samples were active photocatalysts. Other measured water quality parameters of the pre-treated POME for both before and after the ZnO photocatalytic treatment are shown in Table 1. It can be found that the BOD5, turbidity, TSS, colour and oil and grease removal efficiencies were 64.8%, 85.1%, 78.9%, 80.5% and 58.1%, respectively. The final pH of the treated POME was found to decrease from pH 8.65 to pH 8.60.

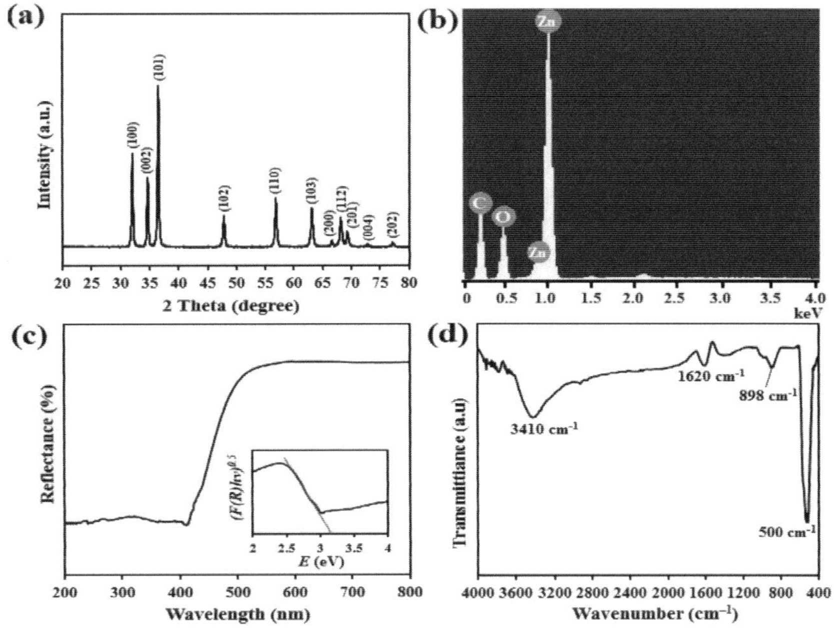

Figure 1. (a) XRD pattern, (b) EDX spectrum, (c) UV-vis DRS spectrum and (d) FTIR spectrum of ZnO micro/nanoflowers. Inset of (c) is the plot of $(F(R)hv)^{0.5}$ vs E (hv).

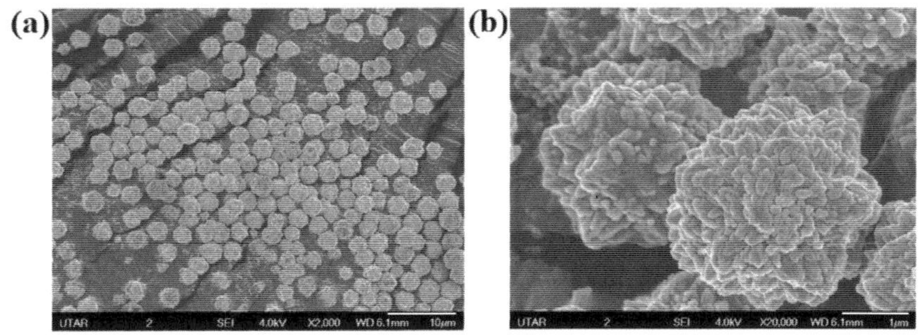

Figure 2. FESEM images of ZnO micro/nanoflowers at different magnifications (a) 2,000x and (b) 20,000x.

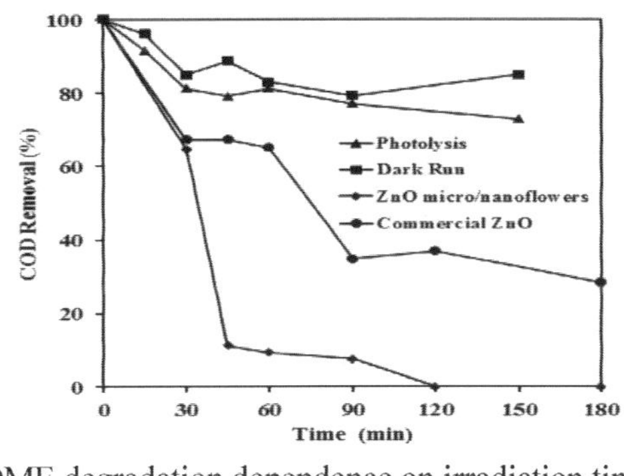

Figure 3. Pre-treated POME degradation dependence on irradiation time at different conditions.

Table 1. Summary of measured pre-treated POME quality parameters before and after ZnO photocatalytic treatment.

Parameter	As-received pre-treated POME	After 120 min UV light	Removal efficiency (%)
COD (mg/L)	50	2	96.0
BOD_5 (mg/L)	54	14	64.8
Turbidity (NTU)	1.75	1.49	85.1
Colour (Pt-Co)	128	27	78.9
TSS (mg/L)	41	8	80.5
Oil and grease (mg/L)	0.43	0.18	58.1
pH	8.65	8.60	–

4. Conclusion

Flower-like ZnO micro/nanostructures were successfully synthesized via a reflux route at low temperature of 65°C. The proposed method was facile and suitable to yield a large fabrication of ZnO micro/nanoflowers. The physico-chemical and optical properties of the as-prepared ZnO micro/nanoflowers were confirmed by XRD, FESEM, EDX, FTIR and UV–vis DRS analyses. The ZnO micro/nanoflowers have displayed good photocatalytic performance for the pre-treated POME degradation under UV irradiation. The removal efficiencies of other water quality parameters such as BOD_5, turbidity, TSS, colour, oil and grease and pH were also obtained.

Acknowledgements
This work was supported by the Universiti Tunku Abdul Rahman (UTARRF/2016–C2/S03 and UTARRF/2017–C1/L02) and Ministry of Higher Education of Malaysia (FRGS/1/2015/ TK02/ UTAR/02/2 and FRGS/1/2016/TK02/UTAR/02/1).

References

[1] Bello M M, Nourouzi M M, Abdullah L C, Choong T S Y, Koay Y S and Keshani S 2013 *J. Hazard. Mater.* **262** 106
[2] Poh P E and Chong M F 2009 *Bioresour. Technol.* **100** 1
[3] Cheng C K, Derahman M R and Khan M R 2015 *J. Environ. Chem. Eng.* **3** 261
[4] Ng K H, Lee C H, Khan M R and Cheng C K 2016 *Chem. Eng. J.* **286** 282
[5] Ng K H and Cheng C K 2016 *Chem. Eng. J.* **300** 127
[6] Lam S M, Sin J C, Abdullah A Z and Mohamed A R 2012 *Desalin. Water Treat.* **41** 131
[7] Tian Y L, Chang L L, Lu J L, Fu J, Xi F N and Dong X P 2013 *ACS Appl. Mater. Interfaces* **5** 7079
[8] APHA, AWWA, WPCF 2005 *21st ed. American Public Health Association*, Washington, D.C., USA
[9] Bakraouy H, Souabia S, Diguaa K, Dkhissia O, Sabarb M and Fadil M 2017 *Proc. Saf. Environ.* **109** 621
[10] Sin J C, Lam S M, Satoshi I, Lee K T and Mohamed A R 2014 *Appl. Catal. B: Environ.* **148–149** 258
[11] Sin J C and Lam S M 2016 *Mater. Lett.* **182** 223
[12] Karnan T and Selvakumar S A S 2016 *J. Mol. Struc.* **1125** 358
[13] He J, Luo Q, Cai Q Z, Li X W and Zhang D Q 2011 *Mater. Chem. Phys.* **129** 242
[14] Lam S M, Sin J C Abdullah A Z and Mohamed A R 2013 *J. Mol. Catal. A: Chem.* **370** 123

A Study on the Relationship between Design Elements of Outdoor Leisure Spaces and Types of Leisure Activities in Sustainable Community Development – A Case Study on Tainan, Taiwan

J H Chang[1] and H Zhang [1,*]

[1]Department of Architecture, National Cheng Kung University, 701 Tainan, Taiwan

* E-mail: changlin@mail.ncku.edu.tw

Abstract. Sustainable community development encompasses three aspects: "Lifestyle", "Production" and "Ecology". Among them, "Lifestyle" is closest to people and reflects basic human needs. Creating outdoor spaces that encourage residents to engage in leisure activities will not only provide them with spiritual sustenance but also fulfil one of the key criteria in sustainable community development. This study explores the relationship between design elements of outdoor leisure spaces and types of leisure activities from residents' perspective with the goal to inform future spatial planning. The study collected 365 valid questionnaires from Tainan residents. Factor analysis was used to extract factors from design elements of outdoor leisure spaces, and regression analysis was applied to understand the effect level. The result shows design elements have positive effect on the types of leisure activities. In addition, different elements exert different influences on the choice of activities.

1.Introduction

As awareness in sustainable development and popularity in leisure and recreational activities grow, what people demand from their public leisure spaces are also changing with the times or their personal needs. Residents enjoy leisure activities both inside their homes and in their surrounding environment [1]. The community is the basic building block of a city. Public leisure spaces in a community not only serve as open space but also as a platform for residents' outdoor living and voluntary daily activities. Leisure activities undertaken by residents in the community are varied [2]. However, living in a crowded city, they also tend to choose certain types of leisure activities under the influence of external environmental factors. It is important to meet their leisure needs while ensuring sustainable development.

2.Design Elements of Community Outdoor Leisure Space and Types of Leisure Activities

Undertaking leisure activities in the community has a positive effect on lifestyle [3].From the perspective of urban planning, the community environment can be spatially subdivided into architecture and town planning, transport links and green space [4]. Kevin Lynch asserts six spatial elements in site planning: paving, street furniture, buildings, open spaces, artifacts and plants [5]. Abbasi argues that different types of space (such as streets, green space, parks, squares, etc.) have effect on our physical, mental health and quality of life [6]. However, studies have shown that main factors affectiing residents' utilization rate of outdoor sapces are distance, attractiveness of space [7] and space maintenance [8]. Providing more and better spaces in the community is a good way to promote physical activities and improve quality of life.

Therefore, we must understand elements of the natural and built environment in the community in order to design better outdoor leisure spaces [9] and attract residents to engage in diversified leisure activities in the outdoors.

The objectives of the study may be summarized as follows: 1. Understand how residents evaluate outdoor leisure spaces. 2. Understand what types of leisure activities resident undertake. 3. Explore the relationship between design elements of outdoor leisure spaces and types of leisure activities.

3.Research Methods

This study explores the relationship between design elements of outdoor leisure spaces and the types of leisure activities from the perspective of residents. We used factor analysis to extract factors from the design elements of outdoor leisure spaces, and then applied regression analysis to understand the effect level.

3.1 Measurement Scale

After reviewing and compiling past literature, the study proposes 6 dimensions and 30 items as a scale for design elements of community outdoor leisure spaces. The dimensions include: "Public Art and Signboard", "Communication Space", "Park", "Passage", "Night Illumination" and "Circulation". The scale for the types of leisure activities uses 4 diminsions: "Social", "Exercise", "Recreation" and "Knowledge and Culture".

3.2 Scope and Subject of Research

Tainan is the cultural and historical hearth of Taiwan, and its residents serve as the subjects of the study. The stratified-quota sampling method was used and a comparison was made with the general population to ensure the profile of variables fits the general population. 377 questionnaires were distributed and 365 valid ones were returned. The study sites are shown in Figure 1.

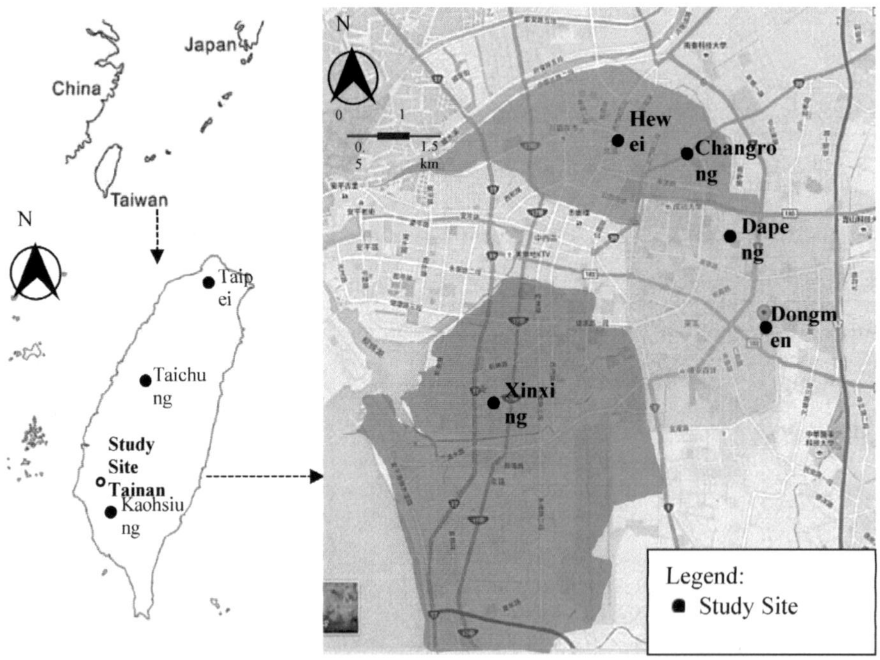

Figure 1. Maps of the study site

4.Research Analysis

4.1 Reliability Analysis

The Cronbach's α for design elements of community outdoor leisure spaces is 0.943, and the Cronbach's α for types of leisure activities is 0.688. According to the reliability standard of Cronbach's α, $\alpha \geq 0.70$ represents high reliability, and $0.70 \leq \alpha \geq 0.6$ represents medium reliability [10]. The Cronbach's α for types of leisure activities is 0.688, meaning the reliability of the questionnaires is acceptable.

4.2 Factor Analysis of Design Elements of Community Outdoor Leisure Spaces

Principal component analysis was used to extract factors and varimax rotation was applied. KMO and Bartlett's test were used to test the appropriateness of factor analysis, and then the derived factors were named. Referencing past literature, the factors were named as follows: "Public Art and Signboard", "Communication Space", "Park", "Passage", "Night Illumination" and "Circulation". Factor analysis of the design elements is shown in Table 1 below.

Table 1. Factor analysis of design elements of community outdoor leisure spaces

Items	F1	F2	F3	F4	F5	F6
	Public Art and Signboard	Communication Space	Passage	Park	Night Illumination	Circulation
A35 Styles of the signboard in communication space are varied.	.790	.114	-.115	.116	.196	.137
A34 Public art installations in the communication space are pleasant.	.786	.170	.015	.255	.090	.100
A7 Public art installations along the passage are pleasant.	.757	.052	.286	.177	.096	.109
A21 Public art installations in the park are pleasant.	.754	.134	.151	.371	.108	.037
A8 Styles of signboards along the passage are varied.	.745	.057	.239	.007	.030	.134
A5 Styles of street furniture along the passage is varied.	.529	.130	.264	-.073	.020	.354
A41 The active crowd in the communication space makes one feel energetic.	-.018	.701	.045	.295	.082	.219
A28 The communication space is accessible.	.260	.700	.193	-.094	.108	.141
A39 Natural illumination of the communication space is adequate.	.037	.686	.241	.026	.242	.097
A30 Maintenance of the communication space is good.	.168	.649	.165	.377	.075	.152
A29 The scale of the communication space is appropriate.	.214	.632	.094	.472	-.014	.190
A36 Paving in the communication space is well maintained.	.187	.425	.384	.338	.209	.054
A9 Paving along the passage is well maintained.	.210	.165	.707	.111	.150	.119
A10 Planting along the passage is diversified.	.135	.146	.698	.115	-.109	.180
A14 The active crowd along the passage makes one feel energetic.	.202	.385	.559	-.071	.157	.161
A11 The passage is well shaded.	.376	-.131	.516	.241	.182	.184
A6 The scale of the passage is appropriate to the surrounding buildings.	.126	.132	.492	.301	.065	.283
A12 Natural illumination along the passage is adequate.	-.026	.291	.458	.163	.332	.289

A15 The park is accessible.	-.145	.291	.440	.250	.253	.035
A17 The park is well maintained.	.190	.107	.139	.665	.333	.103
A16 The scale of the park is appropriate.	.134	.217	.107	.644	-.016	.302
A22 Paving in the park is well maintained.	.277	.181	.338	.556	.267	-.006
A24 The park is well shaded	.252	.124	.175	.548	.305	.142
A26 Night illumination in the park is good.	.106	.129	.077	.231	.823	-.011
A40 Night illumination in the communication space is good.	.214	.256	-.025	.166	.781	-.015
A13 Night illumination along the passage is good.	.097	.035	.270	.051	.762	.189
A1 The passage is accessible.	.088	.197	.182	.145	.087	.794
A2 The scale of the passage is appropriate.	.230	.188	.183	.179	.009	.783
A3 The passage is well maintained.	.253	.169	.222	.133	.094	.686
Eigenvectors	9.949	2.430	2.010	1.516	1.192	1.115
Rotation	4.062	3.230	3.107	2.171	2.626	2.469
Cronbach's alpha	0.880	0.840	0.798	0.759	0.812	0.821
KMO	0.903					
Bartlett's test	0.000					

4.3 Correlation Analysis

Pearson product-moment correlation coefficient was used to probe the correlation level between two variables. The result shows design elements of outdoor leisure spaces and types of leisure activities have significant correlation, where the correlation coefficient is 0.310 (shown in Table 2) indicating moderate correlation (**P<0.01). That means design elements have significant effect on the types of leisure activities, and it is worthwhile to analyse the effect level of the factors.

Table 2. Correlation analysis of design elements of community outdoor leisure spaces and types of leisure activities

		Design Elements of Community Outdoor Leisure Space	**Types of Leisure Activities**
Design Elements of Community Outdoor Leisure Space	Pearson Correlation	1.000	0.310**
	Sig. (2-tailed)		0.000
Types of Leisure Activities	Pearson Correlation	0.310**	
	Sig. (2-tailed)		0.000

**. Correlation is significant at the 0.01 level (2-tailed)

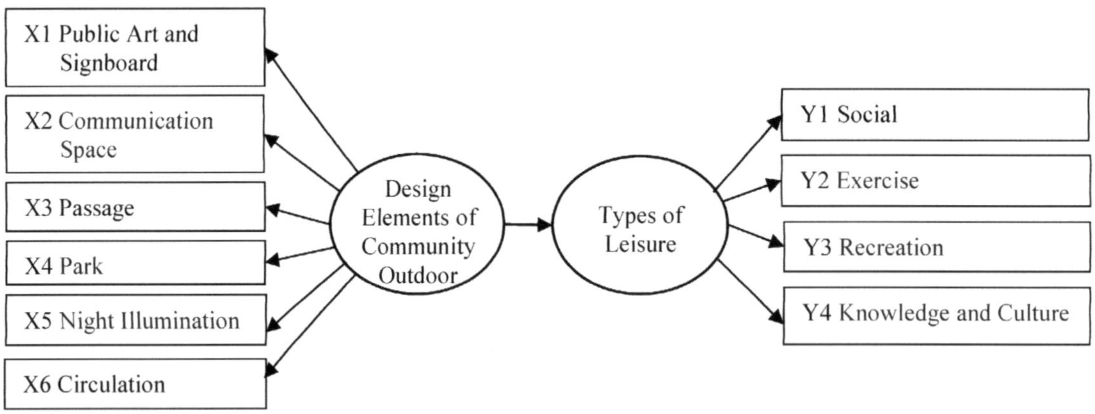

Figure 2. Structure of the study construct

4.4 Multiple Regression Analysis of the Factors of Construct

To address the problem of multi-collinearity, multivariate regression analysis was done first to reduce the error probability of amplification type I. The result showed X1, X2, X3, X4, X5 and X6 have significant effect on Y1, Y2, Y3 and Y4, meaning multiple regression analysis of the factors is feasible. The structure of the study construct is shown in Figure 2.

The types of leisure activities were taken as dependent variables and the factors of design elements of community outdoor leisure spaces as independent ones in the multiple regression analysis (shown in Table 3). The result shows "Public Art and Signboard" and "Communication Space" have significant effect on "Social" and "Communication Space" with medium effect size (0.213); "Public Art and Signboard" has significant effect on "Exercise" (0.145); "Communication Space", "Passage" and "Night Illumination" have significant effect on "Recreation"; "Public Art and Signboard", "Night Illumination" and "Passage" have significant effect on "Knowledge and Culture", and "Public Art and Signboard" has medium effect size (0.264). The hypothesis of the study was confirmed.

Table 3. Regression analysis of the factors of design elements of community outdoor leisure spaces and types of leisure activities

Dependent Variable \ Independent Variable		Public Art and Signboard	Communication Space	Passage	Park	Night Illumination	Circulation	R	R^2	F
Social	Beta	0.094	0.213	-0.026	0.070	0.011	-0.111	0.260	0.068	4.340***
	T	1.461*	2.982**	-0.358	0.928	0.181	-1.681			
Exercise	Beta	0.145	0.138	0.099	0.009	0.004	-0.061	0.288	0.083	5.381***
	T	2.282*	1.938	1.362	0.116	0.073	-0.923			
Recreation	Beta	-0.069	0.184	0.150	-0.022	0.175	-0.115	0.326	0.106	7.084***
	T	-1.092	2.632**	2.085*	-0.298	2.921**	-1.780			
Knowledge and Culture	Beta	0.264	-0.009	0.093	-0.018	0.164	-0.220	0.344	0.118	8.007***
	T	4.238***	-0.132	1.297	-0.251	2.763**	3.429**			

*P<0.05, **P<0.01, ***P<0.001

5.Result and Conclusion

The result shows design elements of community outdoor leisure space not only have positive effect on leisure activities but also influence residents' choice of the activity type. Among the factors of design elements, "Public Art and Signboard" has significant positive effect on "Social", "Exercise" and "Knowledge and Culture"; "Communication Space" has significant positive effect on "Social" and "Recreation"; "Passage" has significant positive effect on "Recreational"; "Night Illumination" has significant positive effect on "Recreation" and "Knowledge and Culture". It is worthwhile to mention that "Circulation" has a significant negative influence on "Knowledge and Culture".

That is, community public art and signboards can be regarded as landmarks and embellishments in spaces, and residents engaged in social, exercise, knowledge and cultural types of activities can more readily feel these elements. Communication space provides a place for residents to stay and communicate. Whether alone or in a group, they need a spot to rest or interact with others, so those engaged in social and recreational leisure activities have stronger feelings for these spaces. A good passage is the only important element for residents engaged in recreational activities. Night illumination not only provides ambient lighting but also becomes an important element in shaping the atmosphere, making residents engaged in recreational and knowledge and cultural activities more sensitive to changes in the environment. The quality of circulation significantly affects the feeling of residents engaged in knowledge and cultural activities. The residents also consider that the higher quality of circulation, the larger volume of traffic, and there will be more external interference, which will reduce the willingness to engage activity in the space. Conversely the lower quality of circulation, the smaller volume of traffic, and there will be less external interference, the possibility of choosing to engage in knowledge and cultural activities here will increase.

Over all, residents with access to a quality outdoor leisure environment tend to engage more in outdoor leisure activities. Creating a quality environment as such requires planners to make field observations and converse with residents and then plan and design suitable spaces that encourage them to participate in outdoor leisure activities during their spare time. Ultimately, residents' needs are met and sustainable development of the environment is made possible.

However, some more questions remain to be answered: how do leisure spaces in different communities satisfy their residents' desire in leisure activities? How do planners promote residents' engagement in outdoor leisure activities to help them relax, adjust the pace of life and face challenges in school or work with renewed motivation? These are issues that need to be further probed.

Acknowledgment

This study was supported by the following Ministry of Science and Technology grant: NSC-103-2511-S-006-001-MY2.

References

[1] Kelly, J. R. (2012). Leisure: Sagamore Publishing.
[2] Cordes, K. A., & Ibrahim, H. M. (1999). Application in Recreation and Leisure. For Today and the Future. NY: McGraw-Hill.
[3] Mehta, V. (2007). Lively streets determining environmental characteristics to support social behavior. Journal of planning education and research, 27(2), 165-187.
[4] Fornara, F., Bonaiuto, M., & Bonnes, M. (2010). Cross-validation of abbreviated perceived residential environment quality (PREQ) and neighborhood attachment (NA) indicators. Environment and Behavior, 42(2), 171-196.
[5] Zhang, H., & Lin, S.-H. (2011). Affective appraisal of residents and visual elements in the neighborhood: A case study in an established suburban community. Landscape and urban planning, 101(1), 11-21.
[6] Abbasi, A., Alalouch, C., & Bramley, G. (2016). Open space quality in deprived urban areas: user perspective and use pattern. Procedia-Social and Behavioral Sciences, 216, 194-205.
[7] Giles-Corti, B., Broomhall, M. H., Knuiman, M., Collins, C., Douglas, K., Ng, K., . . . Donovan, R. J. (2005). Increasing walking: how important is distance to, attractiveness, and size of public open space? American journal of preventive medicine, 28(2), 169-176.
[8] Hur, M., & Nasar, J. L. (2014). Physical upkeep, perceived upkeep, fear of crime and neighborhood satisfaction. Journal of environmental psychology, 38, 186-194.
[9] Kim, J.-H., Lee, C., & Sohn, W. (2016). Urban natural environments, obesity, and health-related quality of life among hispanic children living in inner-city neighborhoods. International journal of environmental research and public health, 13(1), 121.
[10] Nunnally, Jum C. (1967), Psychometric Theory, 1 st ed., New York: McGraw-Hill. (1978), Psychometric Theory, 2d ed., New York: McGraw-Hill.

Atrazine Adsorption Potential of Leonardite from Mae Moh Power Plant

A Rittirat[1], C Chokejaroenrat[2] and A Watcharenwong[1*]

[1] School of Environmental Engineering, Institute of Engineering, Suranaree University of Technology, Nakhon Ratchasima 30000, Thailand.

[2] Department of Environmental Technology and Management, Faculty of Environment, Kasetsart University, Bangkok 10900, Thailand.

*E-mail: w.apichon@sut.ac.th

Abstract. Leonardite is an immature coal from the power plant, which we obtained from Lampang province, Thailand. It is known to contain relatively high humic substance making it suitable as the soil conditioner in agricultural aspect and as an adsorbent. Our objectives were: [1] to determine the leonardite physical and chemical properties (e.g., cation exchange capacity, organic matter content, and electron microscope imaging), and (2) to quantify the adsorption efficiency of atrazine from the soil. We used varied amount of leonardite ranging from 0.05 to 1.60 g to adsorb 8 mg L^{-1} of atrazine from aqueous solution in a batch experiment. This experiment was achieved within 48 hours so that the equilibrium time and isotherm parameters can be determined. The uncarbonized leonardite characteristic revealed that it could adsorb atrazine as its surface possesses mesopore type, relatively high cation exchange capacity (59.89 cmol Kg^{-1}), and high organic matter of 21.62 %. The isotherm study showed the equilibrium time of 24 hours and adsorption isotherm was Langmuir with q_{max} of 3.9683 mg g^{-1}. This study provides proof that leonardite used in the agricultural field can also be useful to adsorb atrazine and simultaneously slow the percolating process of atrazine before contaminating groundwater.

1. Introduction

Leonardite is an immature coal [1]. It is an oxidized form the lignite coal and the deposition of humus [2,3]. Leonardite consists of inorganic and humic substances around 40-80 % [1]. The humic substances in leonardite can be used as a soil conditioner to promote the plant growth. [4]. Humic acid contained organic functional group such as hydroxyl, phenol and carboxyl, its affected for cation exchange in aqueous solution [2]. Moreover, leonardite can be used as the adsorbent for the organic and inorganic substance adsorption's studies. For example, some research group used carbonized-leonardite to remove the congo red in aqueous solution [3]. Some studies related to the removal of As (V) from aqueous solution by leonardite mixing a ferric ion (Fe-leonardite) and the leonardite mixing a clinoptilolite to removal Zn (II) ions from aqueous solution [5]. However, no such feasibility for the adsorption of herbicide as atrazine solution by leonardite have been reported.

Atrazine [2-chloro-4-ethylamino-6-isopropylamino-1,3,5-triazine] is commonly used for a pre- and post-emergent herbicides to control broadleaf weeds. Similarly, atrazine is normally used as herbicides for sugar cane and corn crops. Atrazine has high mobility and can contaminate surface and groundwater by runoff. Moreover, atrazine might be a human carcinogenic compound [6]. The regulation of European Commission No.396/2005 and United State environmental protection agency (US. EPA) has banned atrazine for the agriculture. The maximum atrazine levels in drinking water of these two regulations were at 0.1 ppb and 3 ppb respectively [6, 7]. Nowadays, atrazine has been intensively used in Thailand. Consequently, atrazine removal from the environment has essential importance for the protection of public health and water source.

The aim of this study was to investigate the atrazine adsorption using leonardite. The experiment starts from the characterization of leonardite adsorbent. The characteristic of leonardite such as the elements and components, cation exchange capacity (CEC), organic matter, pore size, and surface area were analyzed. Batch adsorption experiment was conducted to investigate the equilibrium time and batch adsorption isotherm. Throughout the studies, the atrazine solution has been analyzed by high-performance liquid chromatography (HPLC).

2. Methodology

2.1. *Preparation and characterization of leonardite*

The leonardite sample obtained from Mae Moh power plant in Lampang province, Thailand. The leonardite was dried in an oven at 105 °C, 3 hours for moisture removal. Then it was crushed (Gmbh D-42781, Retsch HAAN Germany) and sieved under 150-250 μm (60-100 mesh). Then, the leonardite was kept in a desiccator before use in the experiments. Elements contained in leonardite was characterized by X-ray fluorescence spectrometry (XRF) at 50 KV (XGT-5200, Horiba). The crystalline structure was observed by X-ray diffractometer (XRD) (XRD D2 Phaser, Bruker). The elemental composition (carbon, hydrogen, oxygen, nitrogen, and sulfur) was analyzed by element analyzer (628S/CHN628 LECO). Scanning electron microscope (JSM-6010, JEOL) was used to analyze the morphologies of leonardite. Laser scattering particle size distribution analyzer (LA-950V2, Horiba) was determined surface area and pore size, this method applied by N_2 gas adsorption analysis and degradation at 110 °C for 11 hours.

2.2 *Atrazine adsorption experiments*

Atrazine (Chem service) with a purity of 98.1 % was used as obtained. The stock solution of atrazine 1,000 mg/l was prepared in DI water (18 mΩ).

2.2.1.The equilibrium time of atrazine adsorption. The atrazine concentrations of 1.0, 2.0 and 4.0 mg/l were studies. Leonardite of 0.4 g was added to 25 ml of each atrazine concentration into EPA vials. The vials were shaken in a shaker at 200 rpm. Samples were taken at various contact time at 0-48 hours. Adsorption experiments were performed at room temperature. Then, the samples were measured the atrazine concentrations by HPLC.

2.2.2.Batch adsorption isotherm experiments. Batch adsorption isotherm was using 0.05, 0.1, 0.2, 0.4, 0.8 and 1.6 g of leonardite in 25 ml of 8 mg/l atrazine solutions. Adsorption experiments were performed at room temperature. Samples were collected after 24 hours of adsorption. The samples were determined atrazine concentrations by HPLC.

2.3. Analysis of atrazine concentration by HPLC

The collected sample was spun at 14,000 rpm for 10 min by centrifuge machine. Afterward, the separation of the adsorbent from atrazine solution was performed using nylon filter 0.45 μm. Adsorption experiments were conducted at room temperature. Determinations of atrazine concentration were carried out by HPLC instrument (Agilent 1260 infinite, USA). The UV/Vis detector was set at 220 nm. The column used was a C-18 ACE (5 μm, 4.6 x 250 nm). The mobile phase was acetonitrile: DI water (60:40) and the injection volume was set at 1 ml/min.

3. Results and discussion

3.1. characteristics of Leonardite

The results from chemical characterization of leonardite found that, the pH of leonardite is 2.85 (in H_2O) indicated high acidity. The pH affected the ion exchange capacity on the adsorbent. Ausavasukhi et al.[2] reported that, the efficiency of organic adsorption (congo red; CR) by leonardite was increased at mostly acidity. The organic substance adsorption was decreased at pH increasing. Because the hydroxide ion (OH^-) and carboxyl group (-COOH) on the surface of leonardite can disturb the anionic adsorption of the functional groups of atrazine [2, 8, 9]. Leonardite consists of carbon around 16.54 % and oxygen around 26.67 % as indicated by XRF analysis. The XRD result showed the major phases of leonardite, it has most of the quartz (SiO_2) and gypsum ($CaSO_4 \cdot 2H_2O$). Leonardite has a cation exchange capacity (CEC) of 59.98 %. The high CEC value imply that this material has high organic matter. Hence, it affected the adsorption of atrazine on negative site. Moreover, leonardite consists of humic acid and nutrient which can be used as a soil conditioner. The BET surface area and pore diameter of leonardite were 13.59 m^2/g and 21.65 nm respectively. Leonardite was classified as mesoporous material (2-50 nm) [10]. These pores serve as the path for the pollutant molecule move into a small pore of the particle. The pictures of leonardite were shown in the figure 1.

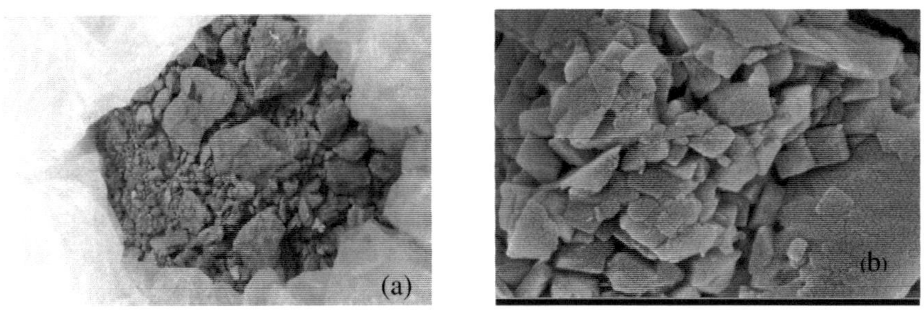

Figure 1. The leonardite at (a) Real picture of leonardite (b) SEM image of leonardite at 50 kx

3.2 Equilibrium time of adsorption

The equilibrium time of the atrazine adsorption on leonardite was around 24 hours. The adsorption efficiencies from 24 hours to 48 hours were similar.

3.3 Adsorption isotherm

The adsorption isotherm studies were conducted with 8 mg/l of atrazine concentration, and the amount of leonardite was varied at 0.05, 0.1, 0.2, 0.4, 0.8 and 1.6 g. All experiments were shaken with 200 rpm at room temperature for 24 hours. The Langmuir isotherm is practical to monolayer adsorption and homogeneous surface. While the Freundlich isotherm is widely for the surface heterogeneous and assumes to multilayer capacity [6]. The linear of Langmuir and Freundlich isotherm was expressed as equation (1) and (2), respectively.

$$1/q_e = 1/(q_{max}K_LC_e) + (1/q_{max}) \quad (1)$$

Where q_{max} is the maximum adsorption capacity (mg/g) and K_L is the Langmuir constant (L/mg), determined from the slope and intercept of linearized plot of $1/q_e$ (q_e is the amount of atrazine adsorbed by leonardite at equilibrium; mg/g) and $1/C_e$ (C_e is concentrations of atrazine at equilibrium in solution; mg/L).

$$\text{Log } q_e = \log K_f + \frac{1}{n}\log C_e \quad (2)$$

Where K_f is the Freundlich adsorption equilibrium constant (mg/g) and n is a constant indicative of adsorption intensity. Moreover, R^2 value has asserted the adsorption isotherm in a linear question. For the results followed a both of Langmuir and Freundlich isotherm were shown in Figure 2. The parameters of adsorption isotherm were shown in table 1.

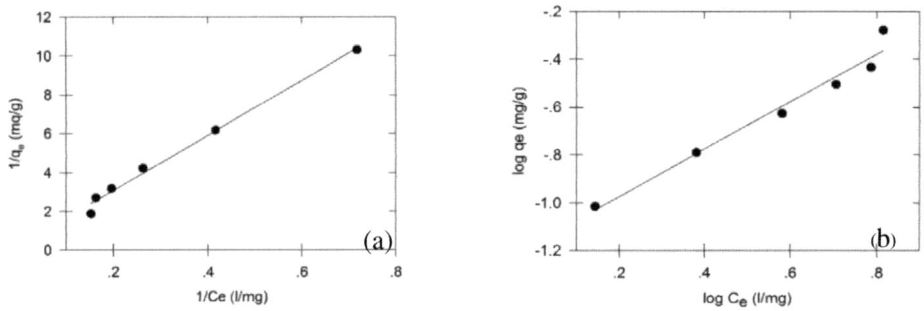

Figure 2. Adsorption isotherm of atrazine on leonardite by (a) Langmuir model and (b) Freundlich model

Table 1. Langmuir and Freundlich isotherm equation for atrazine adsorption by leonardite

Langmuir isotherm				**Freundlich isotherm**			
***E*quation**	R^2	q_{max} (mg/g)	K_d (L/mg)	equation	R^2	n	k_f
Y = 14.1539x + 0.2520	0.9917	3.9683	0.0178	Y = 0.9923x – 1.1723	0.9674	1.0077	0.0673

The atrazine adsorption isotherm by leonardite at equilibrium conditions followed a both of Langmuir and Freundlich isotherm. The maximum Langmuir adsorption capacity is 3.9683 mg/g, which is a monolayer of surface area to limit of molecules to adsorb atrazine. The mechanism of atrazine adsorption may be deepened on the leonardite's surface and it's binding with the functional group of atrazine [9]. Likewise, the atrazine adsorption depended on the pore size on the adsorbent, because the molecular size of atrazine is on around 0.80-0.85 nm [6]. So the mesopore of leonardite (2-50 nm) rendered an atrazine solution pass through into the inner surface of leonardite [7]. Moreover, hydroxyl and a carboxylic group from humic compounds in leonardite play a key role in atrazine adsorption [11]. Therefore, the functional group and surface area affected by atrazine adsorption by leonardite.

4. Conclusion

This study has shown that the leonardite can adsorbed atrazine from aqueous solution. The equilibrium time of adsorption was around 24 hours. The Langmuir adsorption isotherm fitted well with atrazine adsorption by leonardite. The main factors affected atrazine adsorption by leonardite are the pore diameter and specific surface area. Moreover, the humic compounds in leonardite were dominated by the atrazine adsorption. Finally, leonardite could relieve the herbicides contaminate to the surface and groundwater by runoff. Moreover, the utilization of leonardite as the soil conditioner in agriculture area could be possible.

Acknowledgments

The authors are grateful to Suranaree university of technology for financial support. And authors would like to thank the faculty of veterinary technology and department of environmental technology and management, faculty of environment of Kasetsart university for experimental support.

References

[1] Olivella, M.A., Sole, M., Gorchs, R., Lao, C & Delasheras, F. (2011). Geochemical characterization of a spanish leonardite coal. Arch.Min.Sci. 56, 789-804.

[2] Ausavasukhi, A., Kampoosaen, C. & Kengnok, O. (2015). Adsorption characteristics of congo red on carbonized leonardite. Journal of Cleaner Production. 134, 506-514.

[3] Chammui, Y., Sooksamiti, P., Naksata, W., Thiansem, S. & Arqueropanyo O. (2014). Removal of arsenic from aqueous solution b adsorption on leonardite. Chemical Engineering Journal. 240, 202-210.

[4] Yolcu, H., Seker, H., Gullap, M., Lithourgidis, A. & Gunes, A. (2011). Application of cattle manure, zeolite and leonardite improves hay yield and quality of annual ryegrass (*Lolium multiflorum Lam.*) under semiarid condition. AJCS. 5, 926-931.

[5] Znegin, G. (2013). Effective removal of zine from an aqueous solution using Turkish leonardite-clinoptilolite mixture as a sorbent. Environ Earth Sci. 70, 3031-3041.

[6] Rambabu, N., Guzman, C., Soltan, J., & Himabindu, V. (2011). Adsorption characteristics of atrazine on granulated activated carbon and carbon nanotubes. Chem.Eng.Technol. 35, 272-280.

[7] Chingombe P., Saha B. & Wakeman R.J. (2006). Sorption of atrazine on conventional and surface modified activated carbons. Journal of Colloid and Interface Science. 302, 408-416.

[8] Shiri, J., Keshavarzi, A., Kisi, O., Iturraran-Viveros, U., Bagherzadeh, A., Mousavi, R., & Karimi, S. (2017). Modeling soil cation exchange capacity using soil parameters: Assessing the heuristic models. Computers and Electronics in Agriculture.135, 242-251.

[9] Lupul, I., Yperman, J., Carleer, R., & Gryglewicz, G. (2016). Adsorption of atrazine on hemp stem-based activated carbons with different surface chemistry. Adsorption. 21, 489-498.

[10] Mccusker, L.B., Liebau, F., Engelhardt, G. (2001). Nomenclature of structural and compositional characteristics of ordered microporous and mesoporous materials with inorganic hosts. Pure Appl. Chem 73, No.2, 381-394.

[11] Llado J, Lao-Luque C, B.Ruiz, E.Fuente.(2015). Role of activated carbon properties in atrazine and paracetamol adsorption equilibrium and kinetics. Process safety and Environmental Protection. 95, 51-59.

Inhibition of Anaerobic Biological Treatment :A Review

LiHou[1], Dandan Ji[1] and Lihua Zang[1,*]

[1]School of Environmental Science and Engineering, Qilu University of Technology, Jinan, Shandong, China 250353

*zlh@qlu.edu.cn

Abstract. Anaerobic digestion is a method for treating living and industrial wastewater by anaerobic degradation of organic compounds, which can produce biogas (carbon dioxide and methane mixture) and microbial biomass. And biogas as a renewable resource, can replace the use of ore fuel. In the process of anaerobic digestion, the problems of low methane yield and unstable reaction process are often encountered, which limits the widespread use of this technology. Various inhibitors are the main limiting factors for anaerobic digestion. In this paper, the main factors limiting anaerobic digestion are reviewed, and the latest research progress is introduced.

1.Introduction

Anaerobic fermentation has many advantages, such as lower sludge yield, higher energy recovery rate and lower reaction energy consumption. Thermophilic digestion in anaerobic digestion is more stable than mesophilic digestion, and it can destroy viruses, bacteria, and other pathogens in wastewater more thoroughly. Despite these advantages, however, some operational constraints still hinder the widespread use of anaerobic digestion[1]. In anaerobic digestion, microbial anaerobic metabolism pathways are extremely complex, and acid production and methane production are produced by microorganisms respectively. There are great differences in the reaction stage, substrate requirements, growth environment and environmental sensitivity[2]. How to maintain the stable metabolism between these two stages is the main way to solve the problem of anaerobic digestion instability. As inhibitors are ubiquitous in wastewater and anaerobic sludge, they are generally considered to be the major factors responsible for anaerobic digestion. A variety of substances have been reported to inhibit anaerobic digestion[3]. When a substance changes adversely in the microbial population or inhibits the growth of bacteria, it can be identified as an inhibitory substance. Inhibition generally refers to the reduction of methane yield and accumulation of volatile organic acids. This paper mainly introduces the development status of anaerobic digestion, and the summary of various inhibitors that in anaerobic process is to focus on: (1) inhibition mechanism, (2) the inhibition effect factors, (3) the actual problems in the wastewater treatment process.

2.Inhibitors

In the anaerobic process, the decomposition of anaerobic microorganisms using organic matter to produce methane is the main stage of anaerobic reactions. Therefore, in discussing the influencing factors of anaerobic biological treatment, the factors that affect methanogens are generally discussed. The main influencing factors are temperature, pH,presence of other ions ,toxic substances and so on.

2.1 Temperature

The effect of temperature on anaerobic microorganisms is particularly significant; anaerobic bacteria can be divided into thermophilic bacteria (or high temperature bacteria), thermophilic bacteria (medium temperature bacteria); accordingly, anaerobic digestion is divided into: high temperature digestion (55℃ or so) And the medium temperature digestion (35℃ or so); the reaction rate is about 1.5 to 1.9 times, the medium temperature digestion, the gas production rate is higher, but the methane content of the gas is low; when dealing with pathogens and parasite eggs, Sludge, the high temperature digestion can achieve a better hygienic effect, and the dewatering performance of the sludge after digestion is also good[4]. With the development and application of the new anaerobic reactor, the effect of temperature on anaerobic digestion is no longer important The biomass in the new reactor is very large) and can therefore be carried out at room temperature (20 to 25℃) to save energy and operating costs[5].

2.2 pH and alkalinity

pH is the most important factor in the process of anaerobic digestion. Methanogens are very sensitive to pH changes, and are generally considered to have an optimum pH range of 6.8 to 7.2.The methanogenic bacteria will be seriously inhibited, and further lead to the deterioration of the whole anaerobic digestion process; anaerobic system pH value by a variety of factors: influent pH, water quality (organic matter, organic species, etc.)[6]. The biochemical reaction, acid-base balance, and gas-solid solution between the dissolution balance[7]. Anaerobic system is a pH buffer system, mainly controlled by the carbonate system.The system of fatty acid content increase (cumulative) will consume -HCO_3 in general, so that the pH drop[8].But the role of methanogens can not only consume fatty acids, but also produce-HCO_3 cause the system's pH rise[9]. Alkalinity was once considered an important factor in anaerobic digestion. But in fact, its role is to ensure that the anaerobic system has a certain buffer capacity to maintain the appropriate pH. The anaerobic system in the event of acidification. It takes a long time to recover[10].

The interaction between free ammonia, pH and volatile fatty acids may lead to "suppressing steady state", which runs stably, but has lower methane production [11, 12]. Microbial growth The optimum pH control may reduce the toxicity of ammonia [13]. It is reported that the acidification of crab wastewater improves the performance of UASB reactors, such as the lower COD concentration of effluent [14]. During the anaerobic digestion of cattle manure, as the pH is adjusted to 7.2 to 6.8, methane production will also increase up to four to five times as high as possible[15]. During anaerobic digestion of pig manure, the volatile fatty acid reached 320mg/L when pH was 8. Adjusting the pH to 7.4 resulted in the VFAs re-use and reduced VFAs concentration of 20mg/L. The reason for this phenomenon may be due to the inhibition of ammonia inhibition under low pH conditions[16]. More worthy of attention is that both methane and acid-producing microorganisms have their optimum pH. Even though ammonia is at a secure standard, failure to keep the pH within the proper range may cause reactor failure[16,17].

2.3 Presence of other ions

Studies have shown that some metal ions can inhibit the anaerobic digestion process, such as Mg^{2+}, Na^{+} and Ca^{2+}, which is due to the existence of other ions, resulting in the decline of the toxicity of ions[18-20]. Ammonia and sodium show antagonism in anaerobic digestion process, an ion can antagonize another ions generated toxicity. Under normal circumstances, 0.15 M ammonia can lead to a decrease to acetic acid production by 20%, but when adding 0.05 M of Na^{+} in the system, methane production will increase by 5% (without the addition of the inhibitor sample)[21]. The effect of the ion combination is higher than that of single ionic. For example, when Na^{+} and K^{+} or Na^{+} and Mg^{2+} are combined, methane production will increase by 10% compared with the single addition of Na^{+}[22]. It is reported that when NH_4Cl up to 30g/L, adding 10% (w/v) phosphate can stimulate animal manure to produce biogas increased[23]. This stimulating effect of the phosphoric acid nucleic acid is due in part to the fact that the biomass is immobilized on the mineral particles, which prevents the rinsing of biomass from the reactor. Due to the antagonistic effect of phosphate rock (K^{+}, Ca^{2+}, Mg^{2+}) minerals, inhibition of ammonia is also considered partly due. However, inhibition due to more than 50g/L of NH_4Cl is irreversible and cannot be eliminated by the addition of phosphate[24].Although they are essential for microbial growth, they also affect the specific growth rates of other nutrients. Even if the medium stimulates the growth of microorganisms, the excess will slow down the growth rate, and even higher concentrations can cause severe inhibition or toxicity.

2.4 Toxic substances

Sulfates and other sulfur oxides are easily reduced to sulfides during anaerobic digestion[25]. When the soluble sulfide reaches a certain concentration, the anaerobic digestion process is mainly due to the production of methane processes[26]. Some metals such as Fe can remove S^{2-}, or from the system to remove H_2S can reduce the inhibition of sulfide and ammonia nitrogen is anaerobic digestion of the buffer.But the concentration is too high, so it will produce toxic effects on the anaerobic digestion process[27].At the same time, heavy metals will cause damage to the enzyme system of anaerobic bacteria, cyanide is also toxic substances[28-30].Sulfate reducing bacteria can reduce sulfate to sulfide in an anaerobic reactor. Sulfate reducing bacteria can be divided into two groups, one is incomplete oxidation, mainly degrading ethyl lactate and CO_2, the other is complete oxidation and oxidation of acetic acid to CO_2. Sulfate reduction resulted in inhibition of two stages. The main inhibition was due to the competition between SRB for organic matter and inorganic substance, which inhibited the CH_4 production. The toxicity of sulfide to different bacterial communities is the result of secondary inhibition.

3. Engineering significance in industry

3.1. Food processing industry wastewaters

In food processing industry, anaerobic digestion process can be used in vegetable and fruit canning, milk and dairy products, meat and seafood processing and starch sugar production process. The content of organic matter in food processing wastewater, anaerobic digestion treatment is ideal for use. However, due to the presence of inhibitors, the application of this technique may be hindered. Ion of seafood processing wastewater containing high concentration, including Na^+, Cl^- and sulfate etc.. Also the production of high salt wastewater in vegetables, milk dairy processing industry. Contains a lot of fat, protein and carbohydrate in milk production in dairy wastewater. Contain fat, blood, feces and recalcitrant organic substances in meat processing wastes. The degradation of these proteins and lipids resulted in accumulation of ammonia and long-chain fatty acids, thereby inhibiting the activity of anaerobic microorganisms [31]. These wastes also contain certain concentrations of fungicides and disinfectants, which also inhibit the anaerobic digestion reaction[32]. However, the increase of C / N ratio and dilution can reduce the toxicity of these inhibitors in the combined digestion process.

3.2 Papermaking wastes

Papermaking wastewater mainly comes from the industrial production of paper pulp washing and bleaching two production process. When washing, the effluent is called black water. The concentration of pollutants in black water is very high, and contains a lot of fiber, inorganic salts and pigments and other substances. Papermaking wastewater discharged from the bleaching process also contains a large amount of acid and alkali substances called white water, which contains a large number of fibers and fillers and resins added to the production process.Sulfates are produced mainly in sulfite pulping processes[33, 34]. The removal of sulfide can be realized by converting sulfide ions into elemental sulfur by sulfur bacteria. Removal of sulfate from COD and newsprint mill wastewater by two-stage anaerobic aerobic process reported by Chen and Horan. In the process of peeling wastewater, tannin contributed up to 50% of COD. It is well known that they exhibit methane toxicity that is proportional to the degree of polymerization. Many studies show that halogen compounds are toxic to anaerobic microorganism. In the papermaking process, a large amount of halogenated substances are produced in the bleaching process, and the anaerobic microorganisms are toxic to the wastewater. Some organic inhibitors can be microbial degradation[33]. Therefore, it is important to clarify the sources of inhibitors that may be contained in the waste water for subsequent anaerobic biological treatment.

3.3 Weaving industry wastes

The wastewater from weaving industry is mainly derived from washing, bleaching and dyeing processes of fibrous matter. Several laboratory scale studies show the ability of anaerobic / aerobic biological treatment of textile wastewater[35]. Other studies have shown that textile wastewater easily inhibits methane production and COD removal. Textile wastewater may be potential inhibitors of components of dyes, auxiliaries (polyacrylic acid, phosphonic acid), surfactant (alkyl phenol polyoxyethylene ether), chloroform, and heavy metals[36, 37].

4. Conclusion

Anaerobic digestion is a kind of waste treatment technology that uses anaerobic microorganisms to degrade the waste to produce methane. It is mainly used in domestic wastewater and industrial waste.

But solid waste or wastewater may contain substances that inhibit normal anaerobic reactions, and even toxic substances. It is mainly ammonia, sulfide, heavy metals etc.. The accumulation of these inhibitors can lead to abnormal processes in anaerobic digestion, such as decreased methane production or reduced methane content. Due to the different reaction substrates and experimental conditions, the research results of inhibition factors in anaerobic process are different. Therefore, the stable operation of anaerobic reaction can be ensured only when the composition information of anaerobic reaction substrate is known. The present study results show that, with other waste digestion, and in anaerobic reaction before the removal of inhibitory substances in the substrate, can improve the ability of anaerobic microorganisms resistance to toxic substances, so as to improve the efficiency of anaerobic digestion.

5. Acknowledgments

This work was supported by the National Key Technology Research and Development Program of the Ministry of Science and Technology of China [grant numbers 2014BAC25B01, 2014BAC28B01]; the Technology Development Plan of Jiangsu Province, China [grant number BE2015662]; and the Open Subject of Jiangsu Key Laboratory of Anaerobic Biotechnology [grant number JKLAB201605].

References

[1] DUPLA M, CONTE T, BOUVIER J C, et al. Dynamic evaluation of a fixed bed anaerobic digestion process in response to organic overloads and toxicant shock loads [J]. Water Science & Technology A Journal of the International Association on Water Pollution Research, 2004, 49(1): 61-8.

[2] KELLEHER B P, LEAHY J J, HENIHAN A M, et al. Advances in poultry litter disposal technology--a review [J]. Bioresour Technol, 2002, 83(1): 27-36.

[3] DEMIREL B, YENIG N O. Two - phase anaerobic digestion processes: a review [J]. Journal of Chemical Technology & Biotechnology, 2002, 77(7): 743-55.

[4] BOUCQUEY J B, RENARD P, AMERLYNCK P, et al. High-rate continuous biodegradation of concentrated chlorinated aliphatics by a durable enrichment of methanogenic origin under carrier-dependent conditions [J]. Biotechnology & Bioengineering, 1995, 47(3): 298-307.

[5] LLOYD K G, ALPERIN M J, TESKE A. Environmental evidence for net methane production and oxidation in putative ANaerobic MEthanotrophic (ANME) archaea [J]. Environmental Microbiology, 2011, 13(9): 2548-64.

[6] FANG H H, LIU H. Effect of pH on hydrogen production from glucose by a mixed culture [J]. Bioresource Technology, 2002, 82(1): 87.

[7] LI Y, ZHU J, WU X, et al. The effect of pH on continuous biohydrogen production from swine wastewater supplemented with glucose [J]. Applied Biochemistry & Biotechnology, 2010, 162(5): 1286-96.

[8] RAY S, CHOWDHURY N, LALMAN J A, et al. Impact of Initial pH and Linoleic Acid (C18:2) on Hydrogen Production by a Mesophilic Anaerobic Mixed Culture [J]. Journal of Environmental Engineering, 2008, 134(2): 110-7.

[9] PRAKASHAM R S, BRAHMAIAH P, SATHISH T, et al. Fermentative biohydrogen production by mixed anaerobic consortia: Impact of glucose to xylose ratio [J]. International Journal of Hydrogen

Energy, 2009, 34(23): 9354-61.
[10] STAVROPOULOS K P, KOPSAHELIS A, ZAFIRI C, et al. Effect of pH on Continuous Biohydrogen Production from End-of-Life Dairy Products (EoL-DPs) via Dark Fermentation [J]. Waste & Biomass Valorization, 2016, 7(4): 753-64.
[11] DUNFIELD P, KNOWLES R, DUMONT R, et al. Methane production and consumption in temperate and subarctic peat soils: response to temperature and pH [J]. Soil Biology & Biochemistry, 1993, 25(3): 321-6.
[12] BERGMAN I, BO H S, NILSSON M. Regulation of methane production in a Swedish acid mire by pH, temperature and substrate [J]. Soil Biology & Biochemistry, 1998, 30(6): 729-41.
[13] GLATZEL S, STAHR K. Methane and nitrous oxide exchange in differently fertilised grassland in southern Germany [J]. Plant & Soil, 2001, 231(1): 21-35.
[14] HORIUCHI J I, SHIMIZU T, TADA K, et al. Selective production of organic acids in anaerobic acid reactor by pH control [J]. Bioresour Technol, 2002, 82(3): 209-13.
[15] ZOETEMEYER R J, HEUVEL J C V D, COHEN A. pH influence on acidogenic dissimilation of glucose in an anaerobic digestor [J]. Water Research, 1982, 16(3): 303-11.
[16] DAI X, LI X, ZHANG D, et al. Simultaneous enhancement of methane production and methane content in biogas from waste activated sludge and perennial ryegrass anaerobic co-digestion: The effects of pH and C/N ratio [J]. Bioresource Technology, 2016, 216(323-30.
[17] ZHANG C, SU H, BAEYENS J, et al. Reviewing the anaerobic digestion of food waste for biogas production [J]. Renewable & Sustainable Energy Reviews, 2014, 38(5): 383-92.
[18] MCCARTNEY D M, OLESZKIEWICZ J A. Sulfide inhibition of anaerobic degradation of lactate and acetate [J]. Water Research, 1991, 25(2): 203-9.
[19] BRAUN R, HUBER P, MEYRATH J. Ammonia toxicity in liquid piggery manure digestion [J]. Biotechnology Letters, 1981, 3(4): 159-64.
[20] HENDRIKSEN H V, AHRING B K. Effects of ammonia on growth and morphology of thermophilic hydrogen-oxidizing methanogenic bacteria [J]. Fems Microbiology Letters, 1991, 85(3): 241-6.
[21] ZHOU H B, QIU G Z. Inhibitory effect of ammonia nitrogen on specific methanogenic activity of anaerobic granular sludge [J]. Journal of Central south University, 2006, 13(1): 63-7.
[22] SCHNEIDERS D, SILVA J D D, TILL A, et al. Specific Methanogenic Activity (SMA) of industrial sludge from the aerobic and anaerobic biological treatment [J]. Ambiente E Água An Interdisciplinary Journal of Applied Science, 2013, 8(2): 135-45.
[23] SOLON K, FLORES-ALSINA X, MBAMBA C K, et al. Effects of ionic strength and ion pairing on (plant-wide) modelling of anaerobic digestion [J]. Water Research, 2015, 70(235-45.
[24] XAVIER F A, CHRISTIAN K M, KIMBERLY S, et al. A plant-wide aqueous phase chemistry module describing pH variations and ion speciation/pairing in wastewater treatment process models [J]. Water Research, 2015, 85(255-65.
[25] CHEN J L, ORTIZ R, STEELE T W J, et al. Toxicants inhibiting anaerobic digestion: A review [J]. Biotechnology Advances, 2014, 32(8): 1523-34.
[26] DONG C J, PAN Q Y. Inorganic and Organic Toxicants in Anaerobic Digestion of Industrial Wastewater [J]. Environmental Protection of Chemical Industry, 2001,

[27] HICKEY R F, VANDERWIELEN J, SWITZENBAUM M S. The effects of organic toxicants on methane production and hydrogen gas levels during the anaerobic digestion of waste activated sludge [J]. Water Research, 1987, 21(11): 1417-27.

[28] MUDHOO A, KUMAR S. Effects of heavy metals as stress factors on anaerobic digestion processes and biogas production from biomass [J]. International Journal of Environmental Science & Technology, 2013, 10(6): 1383-98.

[29] LI Y, YANG Y, JIANG W, et al. Integrated treatment of municipal sewage sludge by deep dewatering and anaerobic fermentation for biohydrogenproduction [J]. Environmental Science & Pollution Research International, 2015, 22(4): 2599-609.

[30] GUPTA P, AHAMMAD S Z, SREEKRISHNAN T R. Improving the cyanide toxicity tolerance of anaerobic reactor: Microbial interactions and toxin reduction [J]. Journal of Hazardous Materials, 2016, 315(52.

[31] VIDAL G, CARVALHO A, M NDEZ R, et al. Influence of the content in fats and proteins on the anaerobic biodegradability of dairy wastewaters [J]. Bioresource Technology, 2000, 74(3): 231-9.

[32] SALMINEN E A, RINTALA J A. Anaerobic Digestion of Poultry Slaughtering Wastes [J]. Environmental Technology, 1999, 20(1): 21-8.

[33] ALI M, SREEKRISHNAN T R. Aquatic toxicity from pulp and paper mill effluents: a review [J]. Advances in Environmental Research, 2001, 5(2): 175-96.

[34] THOMPSON G, SWAIN J, KAY M, et al. The treatment of pulp and paper mill effluent: a review [J]. Bioresource Technology, 2001, 77(3): 275.

[35] VANDEVIVERE P C, BIANCHI R, VERSTRAETE W. ChemInform Abstract: Treatment and Reuse of Wastewater from the Textile Wet - Processing Industry: Review of Emerging Technologies [J]. Cheminform, 1998, 29(43): no-no.

[36] FEITKENHAUER H. Fast online determination of surfactant inhibition in acidic phase bioreactors [J]. Water Science & Technology A Journal of the International Association on Water Pollution Research, 2004, 49(1): 23-9.

[37] LEE Y H, PAVLOSTATHIS S G. Reuse of Textile Reactive Azo Dyebaths following Biological Decolorization [J]. Water Environment Research, 2004, 76(1): 56-66.

IOP Conf. Series: Earth and Environmental Science **112** (2018) 012007 doi:10.1088/1755-1315/112/1/012007

Microbial Community Composition in a Simultaneous Nitrification and Denitrification Bioreactor for Domestic Wastewater Treatment

Chen Chen[1*], Wukun Ouyang[2], Shan Huang[3*] and Xiaochun Peng[4]
[1]Researcher, South China Institute of Environmental Sciences, Ministry of Environmental Protection, China
[2]Student, the Hun School of Princeton, USA
[3]Research Associate, Princeton University, USA
[4]Researcher, South China Advanced Environmental Technologies Co., LTD, Guangzhou, China

*chenchen7@scies.org

Abstract. Traditional domestic wastewater treatments rely on aerobic processes followed by anaerobic processes. The aerobic step in which ammonium and organic carbon are oxidized, calls for large oxygen input, while the anaerobic process often requires extra carbon input. The challenge of synchronizing both processes is to maintain an active nitrifiers sludge under low dissolved oxygen (DO) condition. In this study, a membrane bioreactor was established and operated stable with low DO of 0.1-0.4 mg L^{-1}. Chemical indicators were determined daily, and bacterial community was checked by qPCR and 16S rDNA sequencing every month. After 2 months incubation, the bioreactor reached to a stable removal rate of total nitrogen around 50% and total organic carbon around 90% with the retaining time of 12 h. The sludge showed enrichment of low DO nitrifiers (*Nitrosomonadaceae*, *Chitinophagaceae*, and *Nitrospiraceae*) which were different from sludge in other regular wastewater treatment plants with aerobic and anaerobic cycles.

1. Introduction

Traditional domestic wastewater treatments rely on an aerobic process followed by an anaerobic process. In aerobic process, organic carbon is oxidized to carbon dioxide and release from water to atmosphere. Meanwhile, ammonium is oxidized to nitrite and then nitrate via nitrification process. The aerobic step calls for certain amount of oxygen input, which turns to energy consumption during wastewater treatment. In the anaerobic process, nitrate is reduced to nitrogen gas (N_2) via denitrification process or anaerobic ammonium oxidization (anammox). Since organic carbon is consumed within aerobic process, the anaerobic process often requires extra carbon input [1, 2].

Therefore, by setting up a bioreactor that combines the aerobic and anaerobic processes, both issues from less oxygen input in the aerobic process and less carbon input in the anaerobic process can be solved.

The challenge of setting up a simultaneous nitrification and denitrification bioreactor is to maintain an active sludge with both active aerobes and anaerobes. Aerobic ammonium oxidizing-bacteria (AOB), nitrite-oxidizing bacteria (NOB), and complete ammonia oxidation (comammox, species *Nitrospira*) are known as the key players in an active aerobic sludge for nitrification [3-7]. Ammonium oxidation archaea (AOA) can also contribute to a high level of ammonium oxidation in the ocean system [8], but they are difficult to be cultured in wastewater treatment system. In an active anaerobic sludge, there are a wide taxonomic diversity of denitrifiers [9]. Another nitrogen removal pathway, anammox, which can use the nitrite came from partial ammonium oxidation by AOB, were usually coupled with AOB and be operated in some wastewater treatment plants recently [10, 11]. All of known anammox bacteria species belongs to *Planctomycetes* phylum perform [12, 13]. In a full-scale wastewater treatment plant, anammox process usually work as a pre-treatment or side-treatment process [11].

A range of operational parameters or wastewater characteristics, such as pH, temperature, dissolved oxygen (DO) and carbon/nitrogen ratio, drive the nitrification and denitrification performance, as well as the microbial communities in wastewater treatment system [14, 15]. Among all these factors, DO is considered to be the key factor and has significant effect on microbial community structure and nitrification rate. For instance, lab-scale bioreactor operated with high and low DO concentration could have different bacterial community structure [16]. Linking the environmental properties and bacterial communities based on 14 wastewater treatment plants, DO concentration significantly shaped microbial community structure of sludge [14].

Thus, the objective of our research is to create a low DO bioreactor to support both nitrifiers and denitrifiers and satisfy the N removal efficiency. The change of microbial community structure under low DO operation is highlighted in this study to better understand the mechanisms of the bioreactor.

2. Materials and Methods

2.1. Sludge collection and bioreactor set-up

Activated sludge samples were combination of a full-scale wastewater treatment system and a lab-scale bioreactor. Both of the systems were operated with same process of anaerobic/anoxic/aerobic (A^2O).

A 3 L membrane bioreactor with a DO probe was established and operated for 6 months. An air pressure was designed to maintain the low DO. When DO was lower than 0.1 mg L^{-1}, the air pressure would open and air could come into the bioreactor. When DO reached 0.4 mg L^{-1}, the air pressure would close and stop the air coming in. The bioreactor maintained at 0.1-0.4 mg L^{-1} DO for 60 days. After that, the bioreactor was maintained at DO with 1-1.2 mg L^{-1} for another 60 days.

To reach a low DO stage, the hydraulic retention time (HRT) of the bioreactor was set as 24 h for 15 days, decreased to 12 h for another 15 days, then further decreased to 6 h for a longer operation situation. The ammonium, total nitrogen (TN) and total organic carbon (TOC) from both influent and effluent were determined every day. The bacterial community was checked by real time quantitative PCR (qPCR) assay and 16S rDNA amplicon sequencing every month.

2.2. *qPCR assay*

Sludge DNA was extracted according to the manual of the Fast DNA spin kit (Bio 101, Qbiogene Inc. CA, USA). The qPCR experiments were carried out using a StepOne Plus (Agilent). Total bacteria were represented via quantification of the 16S rRNA genes, using primer sets 1055f/1392r [17](Table 1). For the detection of denitrifiers, denitrifying functional genes (narG, nirS, nirK, and nosZ,) were quantified with primer sets NarG1/NarG3, NirS3/NirS5, NirK1/NirK5 and NosZ-1f/NosZ-2 [18-21] (Table 1). For the detection of the bacterial *amoA* gene, primer set amoA1f/amoA2r [22] was used. The nitrite oxidizing gene *nrxB* was detected nxrB169F/nxrB638R [23] (Table 1). Anammox bacteria were detected by primer set Amx368f/Amx820r [24] (Table 1). Each PCR mixture (20 μL) was composed of 10 μL of SYBR Premix Ex TaqTM II (2×), 0.8 μL 10 nM of each forward and reverse primers, 0.4 μL ROX Reference Dye II (50×)*3, 6.0 μL ddH2O and 2.0 μL of template DNA (TaKaRa Biotechnology, Japan). The PCR program was initiated with 30 s at 94 °C, followed by 40 cycles of 5 s at 94 °C, 30 s at specific annealing temperatures (Table 1), and 30 s at 70°C. Each assay contained a standard using a serial dilution of plasmids containing specific target genes, independent triplicate templates for each soil sample, and triplicate no template controls (NTC). A melt curve was obtained at the end of each PCR reaction, with the protocol of 1 min at 95°C, 30 s at 55°C and 30 s at 95°C. The single peak of the dissociation curve indicated the specificity of the PCR products.

Table 1. Real time PCR primers used for the amplification in this study.

Primer	Target gene	Sequence (5'-3')	Annealing temperature
1055f	16S rRNA	ATGGCTGTCGTCAGCT	57
1392r	16S rRNA	ACGGGGCGGTGTGTAC	57
NarG1	*narG*	GACAAACTTCGCAGCGG	61
NarG3	*narG*	TCACCCAGGACGCTGTTC	61
NirS3	*nirS*	CCTA(C/T)TGGCCGCC(A/G)CA(A/G)T	57
NirS5	*nirS*	GCCGCCGTC(A/G)TG(A/C/G)AGGA A	57
NirK1	*nirK*	GG(A/C)ATGGT(G/T)CC(C/G)TGGCA	57
NirK5	*nirK*	GCCTCGATCAG(A/G)TT(A/G)TGG	57
NosZ-1f	*nosZ*	TTCGCAATGCCCGAAAGG	57
NosZ-2r	nosZ	AAAACCCCTCTACTTAGTGCCC	57
amoA1f	*AOB amoA*	GGGGTTTCTACTGGTGGT	59
amoA2r	*AOB amoA*	CCCCTCKGSAAAGCCTTCTTC	59
nxrB169F	*nxrB*	TACATGTGGTGGAACA	55
nxrB638R	*nxrB*	CGGTTCTGGTCRATCA	55
Amx368F		TTCGCAATGCCCGAAAGG	57
Amx820R		GGGCACTAAGTAGAGGGGTTTT	57

2.3. *16S rRNA gene amplicon sequencing*

Total DNA of sludge was extracted using CTAB/SDS method. The 16S rRNA gene of V4-V5 was amplified with the barcode by Phusion® High-Fidelity PCR Master Mix (New England Biolabs). PCR products was mixed in equidensity ratios. Then, mixture PCR products was purified with Qiagen Gel Extraction Kit (Qiagen, Germany). Sequencing libraries were generated using TruSeq® DNA PCR-Free Sample Preparation Kit (Illumina, USA) following manufacturer's recommendations and index codes were added. The library quality was assessed on the Qubit@ 2.0 Fluorometer (Thermo

Scientific) and Agilent Bioanalyzer 2100 system. At last, the library was sequenced on an Illumina HiSeq 2500 platform and 250 bp paired-end reads were generated. The 16S rRNA gene amplicon sequencing was run by Novogene Co., China.

2.4. *Data Analyses*

Paired-end reads were assigned to samples based on their unique barcode and truncated by cutting off the barcode and primer sequence. Paired-end reads were split and assembled by FLASH (V1.2.7)[25]. Quality filtering on the raw tags were performed under specific filtering conditions to obtain the high-quality clean tags [26] according to the QIIME(V1.7.0) [27] quality controlled process. The tags were compared with the reference database (Gold database) using UCHIME algorithm (UCHIME Algorithm) [28] to detect chimeric sequences, and then the chimeric sequences were removed [29]. Then the Effective Tags finally obtained.

Sequences analysis were performed by Uparse software (Uparse v7.0.1001) [30]. Sequences with ⩾ 97% similarity were assigned to the same OTUs. Representative sequence for each OTU was screened for further annotation. For each representative sequence, the GreenGene Database [31] was used based on RDP classifier (Version 2.2) [32] algorithm to annotate taxonomic information. In order to study phylogenetic relationship of different OTUs, and the difference of the dominant species in different samples(groups), multiple sequence alignments were conducted using the MUSCLE software (Version 3.8.31) [33]. OTUs abundance information was normalized using a standard of sequence number corresponding to the sample with the least sequences.

Alpha diversity was performed basing on this output normalized data and applied in analyzing complexity of species diversity for a sample through 6 indices, including Observed-species, Chao1, Shannon, Simpson, ACE, Good-coverage. All this indices in our samples were calculated with QIIME (Version 1.7.0) [27] and displayed with R software (Version 2.15.3).

3. Results and Discussion

3.1. *Bioreactor performance and operational conditions*

After 3 months incubation, the performance of the bioreactor was maintained with relatively stable pH about 7, HRT around 12 h, and DO from 0.1 to 0.4 mg L^{-1}. TOC removal efficiency was 86.7%-96.4%, and the average TOC concentration in the effluent was 10.2 mg L^{-1}. TN in the effluent ranged from 35 to 50 mg L^{-1}, while the NH_4^+-N concentration was between 27 and 45 mg L^{-1}. The average removal efficiencies of TN and ammonium were 52.8% and 54.9%, respectively (Figure 1). There were similar removal efficiencies after raising DO to 1 mg L^{-1}. TOC removal was around 83.5%-98.1% and the average TOC concentration in the effluent was 8.2 mg L^{-1}. The average removal efficiencies of TN and ammonium were 47.1% and 50.2%, with the average 38.7 mg L^{-1} TN and 25.2 mg L^{-1} NH_4^+-N in the effluent, respectively.

These results were similar to the other lab-scale bioreactor with A^2O process, HRT around 8 h, biological oxygen demand (BOD) and TN removal rate of 94% and 61%, respectively [15] Full-scale wastewater treatment plant can have more variable influent of chemical oxygen demand (COD) 200-1200 mg L^{-1} and TN 20-120 mg L^{-1} because of operation situation, plant size, surrounding population, processes, etc.[34]. Considering the low DO demand of our bioreactor, the energy input can be much less than those lab-scale bioreactors or wastewater treatment plants.

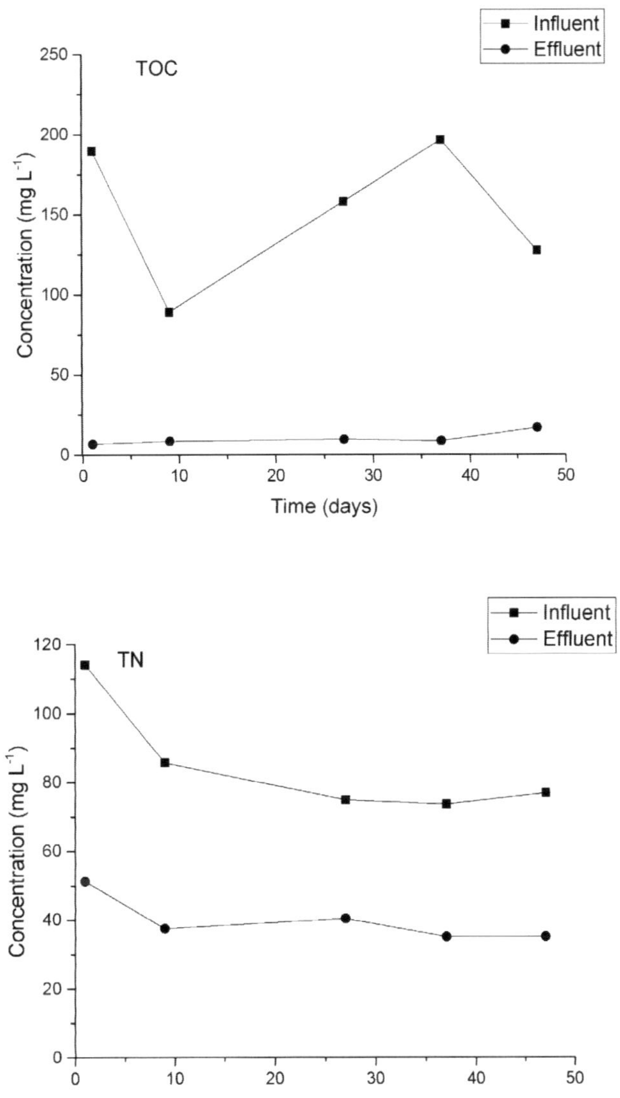

Figure 1. The TOC and TN concentration when bioreactor maintained stable

3.2. Bacterial abundance by qPCR assay

When the bioreactor maintained at low DO of 0.1-0.4 mg L^{-1} for 60 days, the total bacterial abundance (shown by 16S rRNA gene copies) increased from 13.0 to 32.0 copies mL^{-1}. Meanwhile, the functional gene structure had changed. The *nirS, nirK,* and *nxrB* gene copies decreased, while *narG, nosZ,* and *amoA* gene copies increased (Table 2). The number of anammox bacteria was under detection. These results indicated that AOB would take responsible for nitrification during this stage, while nitrite oxidizing process may be inhabited.

When the bioreactor maintained at DO around 1 mg L^{-1}, the total bacterial abundance showed similar increase trend as low DO level. Nevertheless, the bacterial structure changed. Both *amoA* and *nxrB* increased, suggesting that full nitrification process would be enhanced during this stage. In terms of denitrifying genes, *nirS*, *nirK* and *nosZ* showed a decrease trend, while *narG* kept the similar level. These results suggested that denitrification would be inhabited during this stage.

Table 2. Functional gene abundance by qPCR assay

DAY	16S rRNA (× 10^8 copies mL^{-1})	*narG* (× 10^7 copies mL^{-1})	*nirS* (× 10^6 copies mL^{-1})	*nirK* (× 10^6 copies mL^{-1})	*nosZ* (× 10^6 copies mL^{-1})	*amoA* (× 10^6 copies mL^{-1})	*nxrB* (× 10^6 copies mL^{-1})
	DO 0.1-0.4 mg L^{-1}						
0	13.02	2.34	16.34	12.34	3.15	2.75	4.87
20	19.12	2.38	11.23	8.34	8.01	4.10	2.09
40	27.21	6.87	13.32	3.21	9.18	5.51	1.54
60	32.01	6.95	5.64	2.23	9.35	5.90	1.42
	DO 1-1.2 mg L^{-1}						
0	13.24	5.10	6.77	1.98	6.35	1.82	1.23
20	27.45	15.34	22.34	7.34	8.01	4.15	6.87
40	34.34	13.23	23.54	5.34	4.85	5.42	8.31
60	38.54	16.34	14.34	2.13	3.80	5.72	8.09

3.3. Diversity of bacterial communities

After filtering the low quality reads and trimming the adapters, barcodes and primers, there were 46401 effective reads. There sequences were grouped into a total of 411 OTUs, with a 3% sequence dissimilarity cut-off. The number of OTUs was relatively low compared to the number of OTUs within typical bioreactors (2513-3878)[15] or wastewater treatment plants (2176-4123)[34]. This means that our bioreactor had a low bacterial diversity after 6 months incubation. The index of bacterial diversity was shown in Table 3. The abundant OTUs (with > 0.1%) accounted for 18.7% of the total number of OTUs but represented 93.4% of all reads, which mean that some bacterial group would dominate the activated sludge.

Table 3. The diversity of bacterial communities

No.of OTUs	Observed species	Shannon	Simpson	Chao1	ACE
411	411	4.493	0.835	447.0	434.9

At the phylum level, *Proteobacteria* represented the most abundant phylum, accounting for an average of 81.7% of the total sequences. *Bacteroidetes* was the second most abundant phylum (6.5%), followed by *Acidobacteria* (2.7%), *Planctomycetes* (1.2%), *SHA-109* (1.1%) and *Cyanobacteria* (1.1%). Within *Proteobacteria*, *Alphaproteobacteria* (49.9%) and *Betaproteobacteria* (27.9%) were the most dominant subdivision, followed by *Grammaproteobacteria* and *Deltaproteobacteria*. The population of *Bacteroidetes* was mainly composed of two classes: *Sphingobacteria* and *Cytophagia*.

The bacterial diversity at the plylum level and class level showed typical activated sludge communities [7, 14, 15, 34].

At the order level, the bacterial diversity of this study was quite similar as other activated sludge [7, 14, 15, 34]. We also observed an enrichment of *Nitrosomonadales* (top 12, 0.96%) and *Nitrospirales* (top 13, 0.93%) in our bioreactor, indicating an accumulation of nitrifiers after long-term low DO incubation. At the family level, a total of 100 families were observed. This number was quite low compared to typical bioreactors or full-scale wastewater treatment plants, although most of them showed the same families [7, 34]. The abundant (>0.1%) families were illustrated in Figure 2. At the genus level, a total of 140 genera were obtained, compared to usually more than 300 genera in full-scale wastewater treatment plants [7, 14, 34]. The low numbers of order, family and genus of our bioreactor showed that low DO changed the bacterial community structure be to a low diversity but suitable for less oxygen input group.

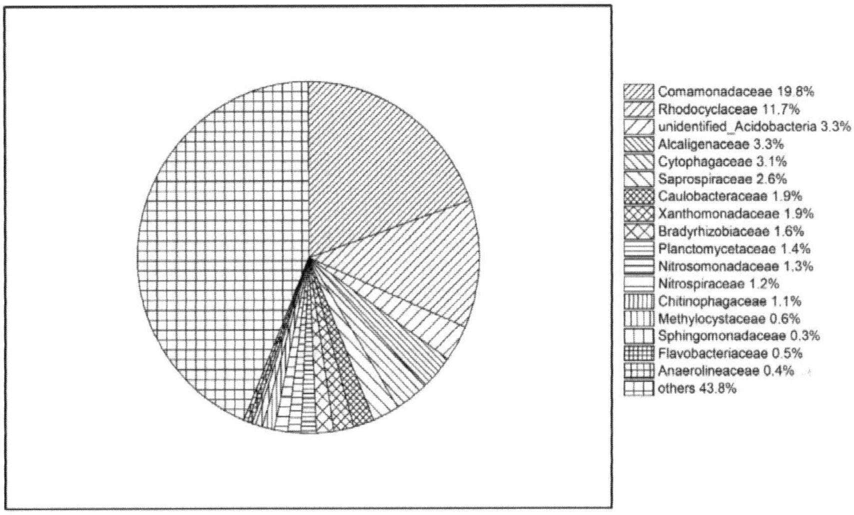

Figure 2. Abundant microbial community of sludge (> 0.1%) at the family level

3.4. *Diversity of N-related bacterial communities*

The N-related bacterial communities has been enriched. Among the abundant (>0.1%) families, N-related bacteria accounted for 51.8% of the total bacterial sequences (Figure 2). Among them, AOB group took part of 5.8%, containing families from *Saprospiraceae*, *Nitrosomonadaceae*, *Chitinophagaceae*, *Sphingomonadaceae*, and *Flavobacteriaceae*. NOB mainly belonged to *Nitrospiraceae* (1.2%). Anammox, which was shown as *Planctomycetaceae*, comprised 1.4%. Denitrifiers (43.4%) came from families including *Comamonadaceae*, *Rhodocyclaceae*, *Alcaligenaceae*, *Cytophagaceae*, *Caulobacteraceae*, *Xanthomonadaceae* and *Bradyrhizobiaceae*.

4. Conclusions

In this study, we set up and operated a bioreactor at a low DO level of 1 mg L^{-1}. After 2 months operation, the bioreactor reached a stable removal rate of TN and TOC around 50% with the HRT of 12 h. The microbial community structure showed enrichment of low DO nitrifiers (*Nitrosomonadaceae*, *Chitinophagaceae*, and *Nitrospiraceae*), denitrifiers, and anammox bacteria groups. These results confirmed that simultaneous nitrification and denitrification were occurring in our bioreactor.

Acknowledgments

Funding of this work was from Program of International S&T Cooperation, the National Key Research and Development Program of China (2016YFE0106600), Chinese National Natural Science Foundation (41501278), Science and Technology Program by Guangdong Financial Department (PM-zx097-201601-024), Science and Technology Program of Guangzhou, China (2016201604030057), and Science and Technology Program of Nansha, China. Dr. Chen Chen was also supported by Scientific Research Foundation for Returned Scholars, Ministry of Environmental Protection of China (PM-zx421-201612-333)

References

[1] Barros P R and Carlsson B 1998 Iterative design of a nitrate controller using an external carbon source in an activated sludge process *Water Science and Technology*. **37**(12) 95-102

[2] Cho J H, Sung S W and Lee I B 2002 Cascade control strategy for external carbon dosage in predenitrifying process *Water Science and Technology*. **45**(4-5) 53-60

[3] van Kessel M A H J, Speth D R, Albertsen M, Nielsen P H, Op den Camp H J M, Kartal B, Jetten M S M and Lucker S 2015 Complete nitrification by a single microorganism *Nature*. **528**(7583) 555–559

[4] Daims H, Lebedeva E V, Pjevac P, Han P, Herbold C, Albertsen M, Jehmlich N, Palatinszky M, Vierheilig J, Bulaev A, Kirkegaard R H, von Bergen M, Rattei T, Bendinger B, Nielsen P H and Wagner M 2015 Complete nitrification by Nitrospira bacteria *Nature*. **528**(7583) 504–509

[5] Daims H, Nielsen J L, Nielsen P H, Schleifer K H and Wagner M 2001 In situ characterization of Nitrospira-like nitrite oxidizing bacteria active in wastewater treatment plants *Applied and Environmental Microbiology*. **67**(11) 5273-5284

[6] Limpiyakorn T, Sonthiphand P, Rongsayamanont C and Polprasert C 2011 Abundance of amoA genes of ammonia-oxidizing archaea and bacteria in activated sludge of full-scale wastewater treatment plants *Bioresource Technology*. **102**(4) 3694-3701

[7] Niu L H, Li Y, Wang P F, Zhang W L, Wang C, Cai W and Wang L Q 2016 Altitude-scale variation in nitrogen-removal bacterial communities from municipal wastewater treatment plants distributed along a 3600-m altitudinal gradient in China *Science of The Total Environment*. **559** 38-44

[8] Konneke M, Bernhard A E, de la Torre J R, Walker C B, Waterbury J B and Stahl D A 2005 Isolation of an autotrophic ammonia-oxidizing marine archaeon *Nature*. **437**(7058) 543-546

[9] Lu, H J , Chandran K and Stensel D 2014 Microbial ecology of denitrification in biological wastewater treatment *Water Research.* **64** 237-254

[10] Quan Z X, Rhee S K, Zuo J E, Yang Y, Bae J W, Park J R, Lee S T and Park Y H 2008 Diversity of ammonium-oxidizing bacteria in a granular sludge anaerobic ammonium-oxidizing (anammox) reactor *Environmental Microbiology.***10**(11) 3130-3139.

[11] Lackner S, Gilbert E M, Vlaeminck S E, Joss A, Horn H and van Loosdrecht M C M 2014 Full-scale partial nitritation/anammox experiences - An application survey *Water Research.* **55** 292-303

[12] Kartal B, Rattray J, van Niftrik L A, van de Vossenberg J, Schmid M C, Webb R I, Schouten S, Fuerst J A, Damsté J S S, Jetten M S M and Strous M. 2007 Candidatus "Anammoxoglobus propionicus" a new propionate oxidizing species of anaerobic ammonium oxidizing bacteria *Systematic and Applied Microbiology.* **30**(1) 39-49

[13] Jetten M S M, van Niftrik L, Strous M, Kartal B, Keltjens J T and Op den Camp H J M 2009 Biochemistry and molecular biology of anammox bacteria *Critical Reviews in Biochemistry and Molecular Biology.* **44**(2-3) 65-84

[14] Wang X H, Hu M, Xia Y, Wen X H and Ding K 2012 Pyrosequencing analysis of bacterial diversity in 14 wastewater treatment systems in China *Applied and Environmental Microbiology.* **78**(19) 7042-7047

[15] Hai R T, Wang Y L, Wang X H, Li Y and Du Z Z 2014 Bacterial community dynamics and taxa-time relationships within two activated sludge bioreactors *Plos One.* **9**(3) 1-8

[16] Park H D and Noguera D R 2004 Evaluating the effect of dissolved oxygen on ammonia-oxidizing bacterial communities in activated sludge *Water Research.* **38**(14-15) 3275-3286

[17] Ryan J G, Estefan G and Rashid A 2011 *Soil and Plant Analysis Laboratory Manual* (Aleppo:International Centre for Agricultural Research in the Dry Areas)

[18] Harms G, Layton A C, Dionisi H M, Gregory I R, Garrett V M, Hawkins S A, Robinson K G and Sayler G S 2003 Real-time PCR quantification of nitrifying bacteria in a municipal wastewater treatment plant *Environmental Science & Technology.* **37**(2) 343-351

[19] Braker G, Fesefeldt A and Witzel K P 1998 Development of PCR primer systems for amplification of nitrite reductase genes (nirK and nirS) to detect denitrifying bacteria in environmental samples *Applied and Environmental Microbiology.* **64**(10) 3769-3775

[20] Scala D J and Kerkhof L J 1998 Nitrous oxide reductase (nosZ) gene-specific PCR primers for detection of denitrifiers and three nosZ genes from marine sediments *Fems Microbiology Letters.* **162**(1) 61-68

[21] Reyna L, Wunderlin D A and Genti-Raimondi S 2010 Identification and quantification of a novel nitrate-reducing community in sediments of Suquia River basin along a nitrate gradient *Environmental Pollution.* **158**(5) 1608-1614

[22] Rotthauwe J H, Witzel K P and Liesack W 1997 The ammonia monooxygenase structural gene amoA as a functional marker: Molecular fine-scale analysis of natural ammonia-oxidizing populations *Applied and Environmental Microbiology.* **63**(12) 4704-4712

[23] Pester M, Maixner F, Berry D, Rattei T, Koch H, Lucker S, Nowka B, Richter A, Spieck E, Lebedeva E, Loy A, Wagner M and Daims H 2014 NxrB encoding the beta subunit of nitrite oxidoreductase as functional and phylogenetic marker for nitrite-oxidizing Nitrospira *Environmental Microbiology.* **16**(10) 3055-3071

[24] Schmid M, Twachtmann U, Klein M, Strous M, Juretschko S, Jetten M, Metzger J W, Schleifer K H and Wagner M 2000 Molecular evidence for genus level diversity of bacteria capable of catalyzing anaerobic ammonium oxidation *Systematic and Applied Microbiology.* **23**(1) 93-106

[25] Magoc T and Salzberg S L 2011 FLASH: fast length adjustment of short reads to improve genome assemblies *Bioinformatics.* **27**(21) 2957-2963

[26] Bokulich N A, Subramanian S, Faith J J, Gevers D, Gordon J I, Knight R, Mills D A and Caporaso J G 2013 Quality-filtering vastly improves diversity estimates from Illumina amplicon sequencing *Nature Methods.* **10**(1) 57-59

[27] Caporaso J G, Kuczynski J, Stombaugh J, Bittinger K, Bushman F D, Costello E K, Fierer N, Pena A G, Goodrich J K, Gordon J I, Huttley G A, Kelley S T, Knights D, Koenig J E, Ley R E, Lozupone C A, McDonald D, Muegge B D, Pirrung M, Reeder J, Sevinsky J R, Tumbaugh P J, Walters W A, Widmann J, Yatsunenko T, Zaneveld J and Knight R 2010 QIIME allows analysis of high-throughput community sequencing data *Nature Methods.* **7**(5) 335-336

[28] Edgar R C, Haas B J, Clemente J C, Quince C and Knight R 2011 UCHIME improves sensitivity and speed of chimera detection *Bioinformatics.* **27**(16) 2194-2200

[29] Haas B J, Gevers D, Earl A M, Feldgarden M, Ward D V, Giannoukos G, Ciulla D, Tabbaa D, Highlander S K, Sodergren E, Methe B, DeSantis T Z, Petrosino J F, Knight R, Birren B W 2011 Chimeric 16S rRNA sequence formation and detection in Sanger and 454-pyrosequenced PCR amplicons *Genome Research.* **21**(3) 494-504

[30] Edgar R C 2013 UPARSE: highly accurate OTU sequences from microbial amplicon reads *Nature Methods.* **10**(10) 996–998

[31] DeSantis T Z, Hugenholtz P, Larsen N, Rojas M, Brodie E L, Keller K, Huber T, Dalevi D, Hu P and Andersen G L 2006 Greengenes, a chimera-checked 16S rRNA gene database and workbench compatible with ARB *Applied and Environmental Microbiology.* **72**(7) 5069-5072

[32] Wang Q, Garrity G M, Tiedje J M and Cole J R2007 Naive Bayesian classifier for rapid assignment of rRNA sequences into the new bacterial taxonomy *Applied and Environmental Microbiology.* **73**(16) 5261-5267

[33] Edgar R C 2004 MUSCLE: multiple sequence alignment with high accuracy and high throughput *Nucleic Acids Research.* **32**(5) 1792-1797

[34] Zhang T, Shao M F and Ye L 2012 454 Pyrosequencing reveals bacterial diversity of activated sludge from 14 sewage treatment plants *Isme Journal.* **6**(6) 1137-1147

Water Quality Assessment of Danjiangkou Reservoir and its Tributaries in China

Linghua Liu[1,*], Wenqi Peng[1], Leixiang Wu[1] and Laisheng Liu[1]

[1] China Institute of Water Resources and Hydropower Research, Beijing, China

* linghualiu@126.com

Abstract. Danjiangkou Reservoir is an important water source for the middle route of the South to North Water Diversion Project in China, and water quality of Danjiangkou Reservoir and its tributaries is crucial for the project. The purpose of this study is to evaluate the water quality of Daniiangkou Reservoir and its tributaries based on Canadian Council of Ministers of the Environment Water Quality Index (CCMEWQI). 22 water quality parameters from 25 sampling sites were analyzed to calculate WQI. The results indicate that water quality in Danjiangkou Reservoir area, Hanjiang River and Danjiang River is excellent. And the seriously polluted tributary rivers were Shending River, Jianghe River, Sihe River, Tianhe River, Jianhe River and Jiangjun River. Water quality parameters that cannot meet the standard limit for drinking water source were fecal coliform bacteria, COD_{cr}, COD_{Mn}, BOD_5, NH_3-N, TP, DO, anionic surfactant and petroleum. Fecal coliform bacteria, TP, ammonia nitrogen, COD_{Mn} were the most common parameters to fail.

1. Introduction

China has been suffering from severe water scarcity problems, especially in northern China. Water availability per capita in China is about 2068 m^3, and 904 m^3 in north China. In order to alleviate the water shortage problems faced in North China, three south to north water transfer projects were proposed. One of the most important projects is the middle route of south to north water transfer project delivering water from the Danjiangkou reservoir to Beijing and Tianjin [1]. The total length of the main channel is 1277km, and annually about 44.8 billion m^3 water was trasfered [2]. the middle water diversion project commenced transferring water in December 2014. From then on, the water quality of Danjiangkou Reservior has attracted considerable public attention.

To evaluate water quality of source water, different water quality assessment methods has been widely studied and used all over the world. Single factor assessment methods were widely used in rivers, reservoirs, and lakes in China. Single factor assessment method is simple and convinient and can be used directly to compare the water quality parameter values with standards. Therefore it is easy to understandard the water quality status with single factor assessment method. However, the single factor assessment method is very strict in some cases. For example, if water quality of all parameters in a drinking water source can meet the standard of drinking water source except faecal coliform, the water body will be evaluated as unfit for drinking purposes using single factor assessment method. However, the water can meet the drinking water quality standard after disinfection treatment in water treatment plant [3].

Water quality index is one of the best water quality evaluation methods [4-5]. Horton [6] first proposed water quality index in 1965, since then, various water quality indices for water quality evaluation have been proposed worldwide [7−13].

However, there is still no universally accepted water quality assessment indices. Most water quality indices in literature or in use in some countries rely on normalizing water quality concentration of parameters with water quality standards and interpretating water 'good' versus 'bad' concentrations [14].

Canadian Council of Ministers of the Environment Water Quality Index (CCMEWQI) compares water quality monitoring results with a benchmark instead of normalizing the measured values with standards, where the benchmark may be a water quality standard or site specific background concentration [15-18]. Many researchers have successfully evaluated water quality with CCMEWQI [19-21].

The objective of this research is to (i) apply the widely used CCMEWQI to evaluate water quality of Danjiangkou Reservoir and 16 major tributaries, using 22 water quality parameters measured on a monthly basis for 16 months; (ii) to identify the main parameters influencing the water quality of Danjiangkou Reservoir and 16 major tributaries.

2. Methodology

2.1. Study area description

Geographical location of Danjiangkou Reservoir is 32°36′ to 33°48′N and 110°59′ to 110°49′E situated in the center of China. It started to impound water in 1967, became functional in 1968 and finished in 1973 with total storage capacity of 17.45 billion m^3 and surface area of 400 km^2 at water level of 157m. The reservoir catchment area is 92,5000 km^2.

From 2005 to 2010, to meet the needs of the South to North Water Diversion Project, the height of Danjiangkou Dam was increased from 162m to 176.6m, and the water level was increased from 157m to 170m Therefore, the Danjiangkou Reservoir is very large with a storage capacity of 29.05 billion m^3, and a water area of 1022.75 km^2.

The Danjiangkou Reservoir dam is located at 0.8km downstream the junction of Danjiang River and Hanjiang River. So the Danjiangkou Reservoir can be subdivided into Han Reservoir and Dan Reservoir. The gate for the water diversion canal is located at Taocha. There are approximately 200 tributaries in the Danjiangkou Reservoir catchment, of which 16 main tributaries make up 90% of the whole drainage area. These 16 tributaries include Hanjiang River, Danjiang River, Qihe River, Laoguan River, Taohe River, Duhe River, Langhe River, Shending River, Sihe River, Guanshan River, Jianhe River, Jianghe River, Quyuan River, Jiangjun River, Tianhe River, Taogou River. Qihe River, Taohe River and Laoguan River are tributaries of Danjiang River. Duhe River, Langhe River, Shending River, Sihe River, Guanshan River, Jianhe River, Jianghe River, Quyuan River, Jiangjun River, Tianhe River and Taogou River are tributaries of Hanjiang River.

The Danjiangkou Reservoir lies within the northern subtropical zone, and the region exhibits a subtropical monsoon climate. The annual mean air temperature is 15-16℃, frost free period is 230-250d and the annual precipitation is 800-1000 mm, 70%-80% of the annual precipitation falls in May to October.

2.2. Water quality monitoring

25 different sampling sites in Danjiangkou Reservoir and 16 main tributaries were selected. Six of these water quality monitoring sites, Hejiawan, Bashang (near the dam), Bashangzhong, Taizishan, Jiangbei Bridge, Taocha (the gate for the water diversion canal), were used to measure the water quality in the reservior. Four of these water quality monitoring sites, Baihe, Liejinba, Yangwei, and Chengjiapo, were used to measure the water quality in Hanjiang River. 11 water quality monitoring sites located near the mouthes of 11 main tributaries of Hanjiang River including Quyuan River, Tianhe River, Duhe River, Jiangjun River, Taogou River, Jianghe River, Sihe River, Guangshan River, Jianhe River, Langhe River, Shending River, were used to measure the water quality in 11 main tributaries of Hanjiang River. The water quality monitoring site Zijingguan was used to measure the water quality of Danjiang River. 3 water quality monitoring sites located near the mouthes of 3 main

tributaries of Danjiang River, Qiihe River, Taogou River, Laoguan River, were used to measure the water quality in tributaries of Danjiang River.

The map of water quality monitoring sites distribution is shown in Figure 1. Water samples were collected in the middle of each month throughout 16 months from July 2013 to October 2014.

Stainless steel water sampler was used to sample water from the sampling location. The sampler was washed with detergent, tap water, 10% hydrochloric acid and tap water, sampling water in turn before use. 22 water quality parameters were analysed on the basis of "Standard Methods for the Examination of Water and Wastewater". Samples for analysing bacteriological parameters were collected in glass bottles cleaned with tap water, detergent, diluted HNO_3, distilled water two times and disinfected at 105 ℃. Samples for analysing physical and chemical parameters were stored in polyethylene bottles. Before sampling, sample bottles were rinsed with sampling water and stored at 4°C to transferred to the laboratory for analysis. Temperature, pH and dissolved oxygen were analysed on site by using thermometer, HACH pH and DO meter respectively. While the other parameters were analysed in the laboratory.

Figure 1. Water Quality Sampling Sites of Danjiangkou Reservoir and its tributaries

2.3. *Water quality standard*

To evaluate water quality of Danjiangkou Reservoir and its tributaries, the Water quality standard of the class III in the *National Environmental Quality Standards for Surface Waters* in China (GB3838-2002) was used. The water quality standard of class III in GB3838-2002 was used for water quality assessment of drinking water source.

2.4. *Calculation of Water Quality Index*

CCMEWQI is calculated with the following equation [14-15].

$$\mathrm{WQI} = 100 - \frac{\sqrt{F_1^2+F_2^2+F_3^2}}{1.732} \qquad (1)$$

Where: F_1 - the percentage of parameters that do not meet water quality standards

$$F_1 = \frac{\text{Number of failed parameters}}{\text{Total number of parameters}} \times 100 \qquad (2)$$

F_2 - the percentage of individual tests that do not meet water quality standards

$$F_2 = \frac{\text{Number of failed tests}}{\text{Total number of tests}} \times 100 \qquad (3)$$

F_3 - the excursion to which the failed test exceed the standard. F_3 is calculated in the following three steps.

1) Excursion calculation of. Excursion is the times of parameter value exceeding the standard. When the parameter concentration must not exceed the standard:

$$\text{Excursion}_i = \frac{\text{Failed parameter value}_i}{\text{Standard}_j} - 1 \qquad (4)$$

When the parameter concentration must not less than the standard:

$$\text{excursion}_i = \frac{\text{Standard}_j}{\text{Failed parameter value}_i} - 1 \qquad (5)$$

2) Calculation of the normalized sum of excursions (nse):

$$\mathrm{nse} = \frac{\sum_{i=1}^{n} \mathrm{excursion}_i}{\mathrm{number\ of\ tests}} \quad (6)$$

3) F_3 is calculated with an equation that scales the normalized sum of the excursions from standards (nse) to yield a value between 0 and 100.

$$F_3 = \frac{\mathrm{nse}}{0.01\mathrm{nse}+0.01} \quad (7)$$

The above equation produces a value of WQI between 0 and 100 and the water quality is ranked into five categories as shown in Table 1.

Table 1. Water quality index categories.

Categories	WQI value
Excellent	95–100
Good	80–94
Fair	60–79
Marginal	45–59
Poor	0–44

3. Results and Discussion

3.1. Descriptive Statistics of 22 Water Quality Parameters

400 samples were taken and 22 parameters were analysed. Among the 22 parameters analysed, 13, 5 and 4 parameters were found having very low concentration lower than method detection limit in the sampling sites in Danjiang Reservoir, Hanjiang River and its tributaries, Danjiang River and its tributaries, respectively. Statistic results for the parameters which include minimum, maximum, average and standard deviation for monitoring site in the Reservoir area, Danjiang River and its tributaries, Hanjiang River and its tributaries, are shown in Table 1, Table 2, and Table 3 respectively. Data less than the method detection limit was counted as the 1/2 method detection limit for statistical calculation.

The pH value of all samples measured varied from 7.12 to 8.7 indicating that the water is slightly alkaline. The maximum and average values of pH did not exceed the limit of Class III in GB3838-2002.

The minimum DO concentration in Danjiangkou reservoir was 5.81mg/L higher than the standard of 5mg/L, which indicates that all DO values in water samples from Danjiangkou Reservoir could meet the Class III water quality standard for drinking purposes. The minimum DO concentration in water samples from tributaries of Danjiang River and Hanjiang River was 1.7 and 0.6mg/L respectively,indicating that Do values sometimes couldn't meet the Class III water quality standard for drinking purposes. The maximum and mean values of DO for the reservoir and rivers were arranged in the following order: Reservoir > Hanjiang River and its tributaries > Danjiang River and its tributaries.

The maximum and average values of all parameters except pH, DO, fluoride, copper for the reservoir and rivers were arranged in the following order: Reservoir < Hanjiang River and its tributaries < Danjiang River and its tributaries. All these phenomena indicate that the pollution status of Hanjiang River and its tributaries is the most serious, and water quality in the reservoir area is the best.

Table 4 shows the statistical results of all tests. 400 samples were taken and 8800 tests were conducted. 96 samples and 2112 tests were used to represent water quality in reservoir area, 64 samples and 1408 tests were used to represent water quality in Hanjiang River, 176 samples and 3872 tests were used to represent water quality in tributaries of Hanjiang River, 16 samples and 352 tests were used to represent water quality in Danjiang River, 64 samples and 1408 tests were used to represent water quality in tributaries of Danjiang river. The failed tests mean the results of the tests cannot meet the standard of Class III in *National Environmental Quality Standards for Surface Waters* (GB3838-2002). Among 8800 tests conducted, 372 tests failed. Water quality in the reservoir area was the best, among 2112 tests with samples from 6 water quality monitoring sites in reservoir, no failed test existed. Among 1408 tests conducted with samples from 4 water quality monitoring sites in

Hanjiang River, namely, Baihe, Liejinba, Yangwei and Chengjiapo, 10 tests were failed. The failed parameter was fecal coliform. Among 3872 tests conducted with samples from water quality monitoring sites in 11 tributaries of Hanjiang River, 345 tests were failed. The water quality in Shending River, Sihe River, Jianghe River, Tianhe Rivre and Jianhe River was poor. The water quality of Laoguan River was also poor based the amount of failed tests.

There were 9 parameters exceeding the water quality standard of Class III in *National* Environmental *Quality Standards for Surface Waters* (GB3838-2002), including fecal coliform, COD_{cr}, COD_{Mn}, BOD_5, NH_3-N, TP, DO, anionic surfactant and petroleum. Figure 2 illustrates the percentage of parameters that exceeded water quality standards. This figure shows which parameters exceeded the standard (F1) and the percentage (%) of each parameter to the total number parameters exceeding water quality standards.

The four most frequently failed parameters were fecal coliform bacteria, TP, ammonia nitrogen, COD_{Mn} with failed frequency of 30%, 14%, 14%, and 14%, respectively (Figure 2). The four parameters accounted for 72% of the exceedances in Hanjiang River, Danjiang River and their tributaries implying that they were the most common parameters exceeding stardards and having the most influence on the water quality index.

Figure 3 shows the frequency of failed tests in different locations. Shending River, Sihe River, Jianghe River, Tianhe River and Jianhe River were the five most seriously polluted rivers with failed frequency of 33.24%, 23.30%, 16.48%, 5.68% and 5.68%, respectively. These five rivers are tributaries of Hanjiang River.

Among the 372 failed tests, 91.4% were in samples from tributaries of Hanjiang River (Figure 4), 4.84% were in samples from the tributaries of Danjiang River, 2.69%, and 1.08% were in samples form Hanjiang River and Danjiang River, respectively. Which indicates that water quality status from good to bad is arranged in the following order: Reservoir > Danjiang River > Hanjiang River > tributaries of Danjiang River > tributaries of Hanjiang River.

Table 1. Statistic Data of Water Quality in the Danjiangkou Reservoir Area (unit: mg/L, except pH and fecal coliform).

No.	Parameter	Min.	Max.	Average	Std. Deviation
1	pH	7.56	8.52	8.12	0.22
2	DO	5.81	12.71	9.40	1.90
3	COD_{Mn}	1.6	2.9	2.0	0.30
4	BOD_5	1.0	2.2	1.5	0.29
5	NH_3-N	0.063	0.349	0.136	0.07
6	TP	0.01	0.02	0.016	0.005
7	F-	0.2	0.28	0.25	0.019
8	Petroleum	<0.01	0.04	0.009	0.008
9	Fecal Coliform(CFU /L)	<20	240	33	46

Table 2. Statistic Data of Water Quality in Danjiang River and itsTributaries (unit: mg/L, except pH and fecal coliform).

No.	Parameter	Min.	Max.	Average	Std. Deviation
1	pH	7.6	8.3	8.0	0.18
2	DO	1.7	12.5	8.81	2.87
3	COD_{Mn}	0.9	6.5	2.10	1.1
4	COD_{Cr}	<10	35	16.37	6.58

5	BOD_5	<0.5	8.0	1.1	1.1
6	NH_3-N	0.027	2.84	0.336	0.560
7	TP	<0.01	0.24	0.06	0.05
8	F^-	0.10	1.08	0.35	0.22
9	Cu	<0.005	0.009	0.007	0.0028
10	Zn	<0.005	0.007	0.027	0.014
11	As	0.0008	0.0062	0.002	0.001
12	Pb	<0.005	0.008	0.008	0.008
13	CN^-	<0.004	0.005	0.0045	0.0007
14	Volatile Phenol	<0.0003	0.0068	0.001	0.001
15	Anionic Surfactant	<0.05	0.124	0.0795	0.022
16	Petroleum	<0.01	0.05	0.034	0.013
17	Fecal Coliform(CFU /L)	<20	130000	10640	24759

Table 3. Statistic Data of Water Quality in Hanjiang River and its Tributaries (unit:mg/L, except pH and fecal coliform).

No.	Parameter	Min.	Max.	Average	Std. Deviation
1	pH	7.12	8.7	7.89	0.37
2	DO	0.6	12.4	8.36	2.4
3	COD_{Mn}	0.6	19.3	3.1	1.95
4	COD_{Cr}	<10	96	22.6	12.05
5	BOD_5	<0.5	44.9	2.8	3.6
6	NH_3-N	0.025	16.1	1.12	1.85
7	TP	<0.01	1.2	0.08	0.14
8	F^-	0.047	0.643	0.201	0.080
9	Cu	<0.005	0.025	0.012	0.009
10	Zn	<0.005	0.111	0.023	0.023
11	As	0.0007	0.0051	0.0017	0.0007
12	Hg	<0.00001	0.00017	0.00001	0.000027
13	Pb	<0.005	0.014	0.0095	0.0064
14	CN^-	<0.004	0.02	0.0067	0.0037
15	Volatile Phenol	<0.0003	0.0306	0.002	0.0033
16	Anionic Surfactant	<0.05	1.4	0.11	0.10
17	Petroleum	0.01	1.22	0.115	0.232
18	Fecal Coliform(CFU /L)	<20	2400000	99610	221255

Table 4. Statistics Data of All Tests.

Location	Sample site	Samples /tests	Failed tests(%)	Failed parameters
Reservoir area	Bahshang	16/352	0	none
	Taocha	16/352	0	none
	Taizishan	16/352	0	none
	Hejiawan	16/352	0	none
	Bashangzhong	16/352	0	none
	Jiangbei Bridge	16/352	0	none
Hanjiang River	Liejinba	16/352	0	none
	Baihe	16/352	1.70%	FC
	Chenjiapo	16/352	0.57%	FC
	Yangwei	16/352	0.57%	FC
tributaries Hanjiang River	Quyuan River	16/352	2.84%	FC
	Jiangjun River	16/352	1.42%	FC
	Taogou River	16/352	0.57%	FC
	Langhe River	16/352	1.14%	FC, COD_{Cr}
	Guanshan River	16/352	1.99%	FC, COD_{Cr}, TP
	Duhe River	16/352	2.56%	FC, COD_{Cr}, NH_3-N
	Jianhe River	16/352	5.68%	FC, COD_{Cr}, BOD_5, COD_{Mn}, NH_3-N, TP
	Tianhe River	16/352	6.82%	FC, COD_{Cr}, BOD_5, COD_{Mn}, NH_3-N, TP
	Sihe Rver	16/352	23.30%	FC, COD_{Cr}, BOD_5, COD_{Mn}, NH_3-N, TP, DO
	Jianghe River	16/352	16.48%	FC, COD_{Cr}, BOD_5, COD_{Mn}, NH_3-N, TP, DO, AS
	Shending River	16/352	33.24%	FC, COD_{cr}, COD_{Mn}, BOD_5, NH_3-N, TP, DO, AS, Petroleum
Danjiang River	Zijingguan	16/352	1.14%	FC, COD_{Mn}, NH_3-N, TP
tributaries of Danjiang River	Qihe River	16/352	0	none
	Taohe River	16/352	0	none
	Laoguan River	16/352	5.11%	FC, COD_{Cr}, COD_{Mn}, BOD_5, NH_3-N, DO
Total		400/8800	4.23%	

FC-Fecal Coliform, AS- Anionic Surfactant

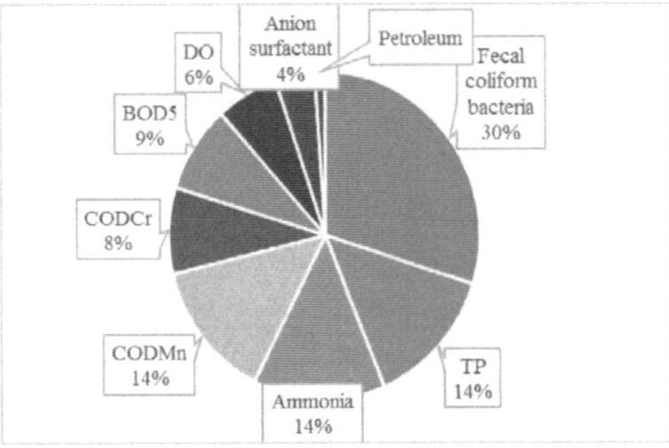

Figure 2. Parameters exceeding the standard

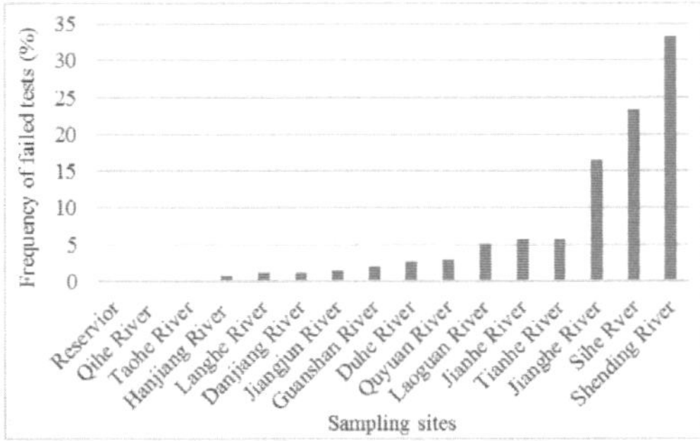

Figure 3. Percentage of failed tests in all tests versus locations

3.2. *Water quality index variation*

Applying the CCMEWQI equation on water quality analysis, Water quality indexes of all monitoring sites in Danjiangkou Reservoir were ranked 100, indicating water quality of the Reservoir was excellent and was protected with absence of threat. Water quality can meet the standard of drinking water source.

Water quality indexes at 4 monitoring sites in Hanjiang River, Chenjiapo, Yangwei, Baihe and Liejingba, were ranked 97.83, 97.85, 97.71 and 100 respectively. Water quality could satisfy the goal for drinking water sources.

Water quality indices for 11 tributaries of Hanjiang River are shown in Table 5. Water quality indices of Taogou River and Langhe River were between 95 and 100, indicating the water quality was excellent. Water quality indexes of Guanshan River, Quyuan River and Duhe River were between 80 and 94, indicating the water quality was good. Water quality indexes of Sihe River, Tianhe River, Jianhe River and Jiangjun River were between 60 and 79, water quality is fair. Water quality index of Jianghe River is 58, water quality is marginal. Water quality index of Shending River is 43.66, Water quality is poor.

Table 5. Water Quality Index on Different Sampling Sites in Hanjiang River Tributaries.

No.	Sampling sites	WQI	Water Quality
1	Taogou River	97.83	Excellent water
2	Langhe River	95.68	Excellent water
3	Guanshan River	93.47	Good water
4	Quyuan River	91.49	Good water
5	Duhe River	90.6	Good water
6	Jiangjun River	77.88	Fair water
7	Jianhe River	75.06	Fair water
8	Tianhe River	64.00	Fair water
9	Sihe River	60.05	Fair water
10	Jianghe River	58.00	Marginal water
11	Shending River	43.66	Poor water

Water quality indexes of Danjiang River and its tributaries are shown in Table 6. Water quality index of Laoguan River was ranked 89.73, water quality was good. Water quality indexes of other sampling sites were higher than 94, water quality was excellent.

Table 6. Water Quality Index on Different Sampling Sites of Danjiang River and its Tributaries.

Site No.	Sampling sites	WQI	Water Quality
1	Zijingguan	95.43	Excellent water
2	Qihe River	100	Excellent water
3	Taohe River	100	Excellent water
4	Laoguan River	89.73	Good water

The percentage of each WQI category of 25 sampling sites is shown in Figure 5. Water quality of 60% of the sampling sites was categorized as excellent water.

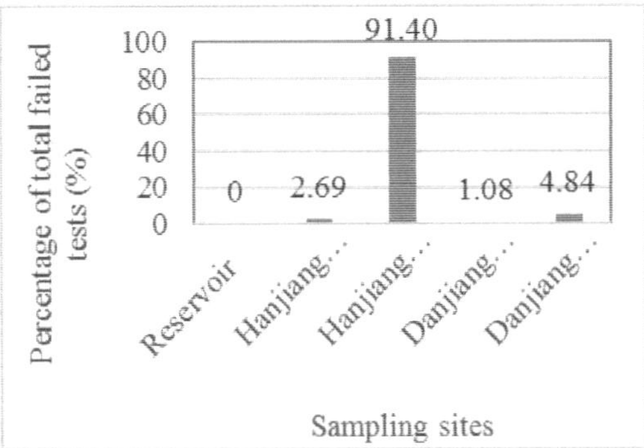

Figure 4. Percentage of total failed tests versus locations

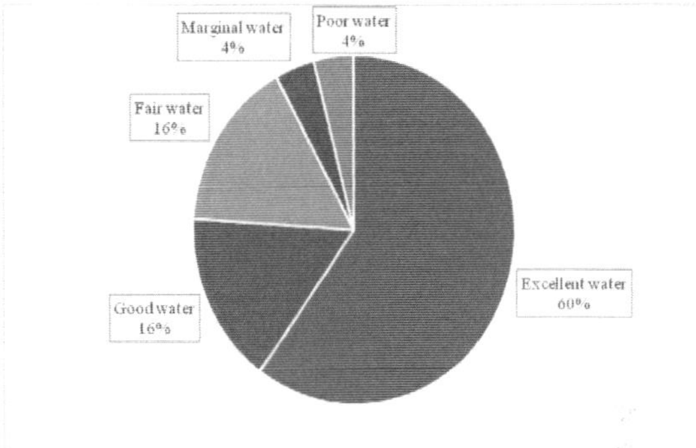

Figure 5. WQI Classification of 25 Sampling Sites

4. Conclusion

Water quality indexes of Danjiangkou Reservoir and its tributaries were calculated by using 22 water quality parameters from 25 sampling sites to evaluate the suitability as drinking water source. The following conclusions can be drawn on the basis of descriptive analysis and water quality index calculation.

1) Water quality indexes of Danjiangkou Reservoir were 100 which indicates water quality in the reservoir area was excellent and the water source was protected with a virtual absence of threat; the water source conditions were very close to natural levels.

2) Water quality indexes of the two major tributaries of Danjiangkou Reservoir, Hanjiang River and Danjiang River, were in the range from 95.43 to 100, indicating water quality of Hanjiang River and Danjiang River is excellent. Only a few tests of water quality monitoring are failed. The failed variables were fecal coliform bacteria, COD_{Mn}, NH_3-N, TP.

3) Water quality indexes of 11 tributaries of Hanjiang River were in the range from 43.66-97.83. Water quality of 11 tributaries from good to bad is in the following order: Taogou River > Langhe River > Guanshan River > Quyuan River > Duhe River > Jiangjun River > Jianhe River> Tianhe River> Sihe River> Jianghe River> Shending River. The seriously polluted tributary rivers were Shending River, Jianghe River, Sihe River, Tianhe River, Jianhe River and Jiangjun River. The failed variables were fecal coliform bacteria, COD_{cr}, COD_{Mn}, BOD_5, NH_3-N, TP, DO, anionic surfactant, petroleum.

4) Fecal coliform bacteria, TP, ammonia nitrogen, COD_{Mn} were the most common parameters to fail and having the most influence on the Water Quality Index.

Acknowledgments
This project *was* supported by International Science and Technology Cooperation Program of China (2015DFG71210) and major science and technology program for water pollution control and treatment (2013ZX07104004)

References

[1] Liu C M and Zhang H. Z., "South-to-north Water Transfer Schemes for China," 2002 Water Resources Development 18 453.

[2] Yang Y, Lei Y, Zhang Q Z, "Quantity versus quality in China's South-to-North Water Diversion Project: a system dynamics analysis" 2015 *Water* 7 2142.

[3] Yin H L and Xu Z X, "Discussion on China's Single- factor Water Quality Assessment Method" 2008 *Water Purification Technology* 27 1(in Chinese)

[4] Simoes F, Moreira A B, Bisinoti M C, Gimeneaz S and Santos M, "Water quality index and dissolved oxygen deficit simple indicators of watersheds pollution" 2007 *Ecological Indicators,* 34 315.

[5] Liou S M, Lo S L and Hu C Y, "Application of two stages fuzzy set theory to river quality evaluation in Taiwan" 2003 *Water Research*, 37 1406.

[6] Horton R K, "An index number system for rating water quality" 1965 *Journal of Water Pollution Control Federation*, 37 300.

[7] House M A and Newsome D H, "Water quality indices for the management of surface water quality" 1989 *Water Science and Technology,* 21 1137.

[8] Brown R M, Mccelland N I, Deininger R A, Tozer R G, "A water quality index – do we dare" 1970 *Water Sewage Works*, 117 339.

[9] Cude C, "Oregon water quality index: A tool for evaluating water quality management effectiveness" 2001 *Journal of the American Water Resources Association* 37 125.

[10] Sargaonkar A, Deshpande V, "Development of an overall index of pollution for surface water based on a general classification scheme in Indian context," 2003 *Environmental Monitoring and Assessment* 89 43.

[11] Liou S M, Lo S L, Wang S H, "A generalized water quality index for Taiwan" 2004 *Environmental Monitoring and Assessment*" 96 35.

[12] Kim A G and Cardone C R, "Scatterscore: a reconnaissance method to evaluate changes in water quality" 2005 *Environmental Monitoring and Assessment* 111 277.

[13] Tsegaye T, Sheppard D, Islam K R, Johnson A, Tadesse W, Atalay A, Marzen L. "Development of chemical index as a measure of in-stream water quality in response to land use and land cover changes" 2006 *Water, Air and Soil Pollution* 174 161.

[14] UNEP/GEMS, Global dinking water index development and sensitivity analysis report., vailable online: http://www.un.org/waterforlifedecade/pdf/global_drinking_water_quality_index.pdf, 2007.

[15] Canadian Council of Ministers of the Environment(CCME). Canadian environmental sustainability indicators, freshwater quality indicator: data sources and methods. Available online: http://www.statcan.ca/bsolc/english/bsolc?catno=16-256-XIE#formatdisp, 2005.

[16] Khan F, Husain T and Lumb A, "Water quality evaluation and trend analysis in selected watersheds of the Atlantic region of Canada," 2003 *Environmental Monitoring and Assessment*, 88 221.

[17] Khan A A, Tobin A, Paterson R, Khan H and Warren R, "Application of CCME procedures for deriving site-specific water quality guidelines for the CCME water quality index" 2005 *Water Quality Research Journal of Canada*, 40 448.

[18] Lumb A, Halliwell D, and Sharma T, "Application of CCME water quality index to monitor water quality: A case of the Mackenzie River Basin Canada" 2006 *Environmental Monitoring and Assessment* 113 411.

[19] Alexakis D, Tsihrintzis V A, Tsakiris G, and Gikas G D, "Suitability of water quality indices for application in lakes in the Mediterranean" 2016 *Water Resources Management*, 30 1621.

[20] Gyamfi C, Boakye R, Awuah E and Anyemedu F, "Application of the CCMEWQI model in assessing the water quality of the Aboabo River, Kumasi-Ghana" 2016 *Journal of sustainable Development* 6 1.

[21] Sharma D and Kansal A, "Water quality ananlysis of River Yamuna using water quality index in the national capital territory India (2000-2009)" 2011 *Applied water science*, Vol.1, 2011 pp.147-157.

Investigations on the Synthesis and Properties of Fe_2O_3 /$Bi_2O_2CO_3$ in the Photocatalytic and Fenton-like Process

Dongxue Sun[1], Tingting Shen[1], Jing Sun[1], Chen Wang[1] and Xikui Wang[2]*

[1]College of Environmental Science and Engineering, Qilu University of Technology, Jinan, Shandong, 250353, P.R. China,
[2] College of Environmental Science and Engineering, Shandong Agriculture and Engineering University, Jinan, Shandong, 251100, P.R. China.

*xk_wang@qlu.edu.cn

Abstract. Catalyst of $Bi_2O_2CO_3$ and Fe_2O_3 modified $Bi_2O_2CO_3$ ($Fe_2O_3/Bi_2O_2CO_3$) were prepared by hydrothermal method and characterized by X-ray diffractions (XRD), scanning electron microscopy (SEM), transmission electron microscope (TEM) and UV-vis DRS. The catalytic activity of $Bi_2O_2CO_3$ and $Fe_2O_3/Bi_2O_2CO_3$ were comparatively investigated in the photodegradation and Fento-like process. Rhodamine B(RhB) was selected as the target pollutant under the irradiation of 300 W xenon lamp. The results indicated that Fe_2O_3 plays a great role in the enhancing the treatment efficiency and the and the maximum reaction rate was achieved at the Fe_2O_3 loading of 1.5%. The Fenton-like degradation rate constant of RhB with bare $Bi_2O_2CO_3$ in dark is 0.4 min^{-1}, while that with 1.5 $Fe_2O_3/Bi_2O_2CO_3$ increases to 28.4 min^{-1} under visible light irradiation, a 71-fold improvement. It is expected to shed a new light for the constructing novel composite photocatalyst and also provide a potential method for the removal of dyes in the aqueous system.

1. Introduction

$Bi_2O_2CO_3$ is a typical Sillén-phase compound, in which the Bi-O atoms layer and CO_3^{2-} layer is orthogonal staggered symbiosis [1]. This structure can improve the separation efficiency of electron-hole pairs [2]. However,with a low absorption of UV light(about 5% of the solar light [3]), the application of $Bi_2O_2CO_3$ photocatalytic materials is limited.Recently, many methods have been employed to overcome its limitations, including heterojunction fabrication. Tian et al. [4] compounded Ag-C_3N_4 and $Bi_2O_2CO_3$ and studied its photocatalytic performance. Wang et al. [3] successfully prepared the composite catalyst of $Bi_2O_2CO_3$/g-C_3N_4 for the first time. This kind of catalyst exhibits high photocatalytic activity on NO in air degradation. Puttaswamy Madhusudan [5] compounded $BiVO_4$ and $Bi_2O_2CO_3$,which can significantly improve its performance. Lin et al. [6] synthesized $Bi_2O_2CO_3$ and Bi_2MoO_6 via a one-pot hydrothermal method. Good photocatalytic activity can be attributed to the heterojunction interface which facilitates the separation of photogenerated electron–hole pairs. Zai et al. [7] researched I-doped $Bi_2O_2CO_3$ and the catalyst exhibits high photocatalytic activity. Lv et al. [8] prepared $Bi_2O_2CO_3/TiO_2$ composite powders which showed that the composite powders have a high photocatalytic activity than TiO_2. Futhermore, Fe_2O_3 [9], a photocatalyst, has also been studied.

AOPs(advanced oxidation processes) has an efficient oxidization ability for organic pollutants, has become an important research task in the field of environmental protection [10], in which Fenton reaction as a fast-growing AOPs technology[11]. Normal Fenton reaction uses the chain reaction between bivalent iron ion (Fe^{2+}) and hydrogen peroxide (H_2O_2) to generate strong oxidation ability and high electronegativity of hydroxyl radicals, most of the organic material in water can be oxidized. Its harsh conditions greatly limits the application in organic wastewater treatment(pH<3). Studies found that photo-Fenton system generated active ferric iron ligand which can regenerate ferrous iron ion through the transfer between ligand-metal charge under visible light. Therefore, finding visible light response catalyst which can efficiently use the solar energy light becomes a new direction in visible light photo-Fenton research.

In this paper, composite catalyst $Fe_2O_3/Bi_2O_2CO_3$ was synthesized by a one-pot hydrothermal method. Especially,N, N-dimethylformamide (DMF) as a carbon source which was used for synthetic $Bi_2O_2CO_3$ was uesd for the first time. The prepared samples were characterized and utilized as a kind of catalyst for the degradation of organic dye under Fenton synergistic effect. The research shows that it is an alternative technology for organic pollutants in wastewater treatment and has broad application prospects.

2. Experimental section

2.1. Preparation of $Bi_2O_2CO_3$ and $Fe_2O_3/Bi_2O_2CO_3$ photocatalysts

All the chemicals were used as received without any further purification. To obtain $Bi_2O_2CO_3/Fe_2O_3$ samples,$Bi(NO_3)_3\cdot 5H_2O$(2 mmol) was dissolved in 30 mL diluted HNO_3(2 M) and stirred for 30 minutes. Then Fe(NO3)3·$9H_2O$ was added into the above solution. The mixed solution was also stirred for anothor 4 hours. Thirdly, the pH was adjusted to 7.0 using ammonia solution under stirring. The fixed solution should be stirred for another 2 hours. With the increasing of the quality of $Fe(NO_3)_3\cdot 9H_2O$, the color of the final precursor was brown deepen. Fourthly, 5 ml DMF was added to above solution. Then the precursor was transferred into a 60 mL Teflon autoclave and maintained at 200℃ for 10 hours.Subsequently,after cooling to room temperature, the products was washed with absolute ethanol and distilled water for several times, then freeze-drying . Especially, The preparation of pure $Bi_2O_2CO_3$ is also described as above, except that it is not necessary to add $Fe(NO_3)_3\cdot 9H_2O$. The final products were described according to the molar percent of $Fe(NO_3)_3\cdot 9H_2O$ used, the as-prepared catalysts were marked as $Bi_2O_2CO_3$, 0.1 $Fe_2O_3/Bi_2O_2CO_3$, 0.5 $Fe_2O_3/Bi_2O_2CO_3$, 1.0 $Fe_2O_3/Bi_2O_2CO_3$, 1.5 $Fe_2O_3/Bi_2O_2CO_3$, 2.0 $Fe_2O_3/Bi_2O_2CO_3$.

2.2. Materials characterization

The catalyst was characterized via powder X-ray diffraction (XRD) with Cu Ka radiation(λ= 1.5404Å) at a scan rate of 0.05° 2θ/s. The morphology of the samples were examined by scanning electron microscope(SEM:Hitach, Japan) and transmission electron microscope(TEM:JEOL, Japan). UV-Vis spectra was achieved by UV-Vis DRS spectrophotometer (U-3900H, Hitachi, Japan) over a range of 100~800 nm using $BaSO_4$ as a reference.

2.3. Photocatalytic activity measurements

The catalysts were evaluated by the degradation of RhB at room temperature. The visible light source was a 300 W Xenon lamp with a cutoff filter(λ<420 nm). Before irradiation,the suspension was magnetically stirred in the dark for 30 min to achieve adsorption/desorption equilibrium. 0.075 g catalyst was mixed with 1mmol/L of H_2O_2. 150 ml RhB solution(RhB of 50 mg/L, pH 3.0) was mixed with the above mixture of catalyst and hydrogen peroxide. 3 ml of solution was taken every five minutes to test. The degradation of RhB was monitored by UV-Vis spectrophotometer.

3. Results and discussion

3.1. Crystal structure analysis

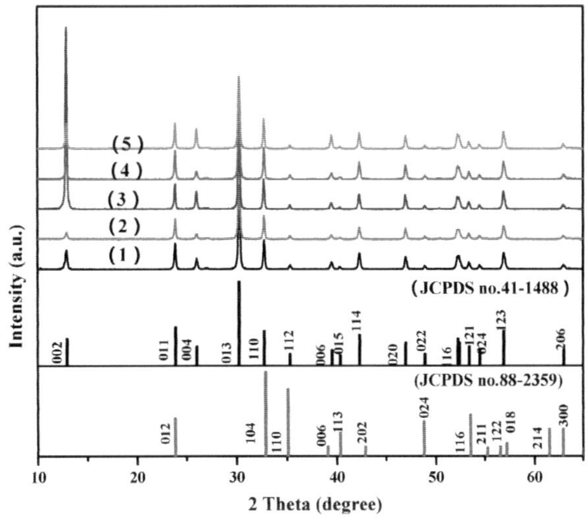

Figure 1. XRD patterns of $Bi_2O_2CO_3$(1),0.1 $Fe_2O_3/Bi_2O_2CO_3$(2),1.0 $Fe_2O_3/Bi_2O_2CO_3$(3),1.5 $Fe_2O_3/Bi_2O_2CO_3$(4),2.0 $Fe_2O_3/Bi_2O_2CO_3$(5) samples.

Figure 1 shows the composites of $Bi_2O_2CO_3$(JCPDS card no.41-1488) and Fe_2O_3(JCPDS card no.88-2 359). It is clear that the prepared catalyst is tetragonal structure. No other diffraction peaks were obser ved, and the sharp and intense peaks indicate a good crystalline nature. The lattice fringes of the comp osite catalysts match well, which can be confirmed by the overlapped peaks between Fe_2O_3 at 2 3 . 8 5°(012), 32.86°(104), 35.08°(110), 39.08°(006), 40.32°(113), 42.87°(202), 48.83°(024), 53.51°(116), 55.25°(211), 56.54°(122) and $Bi_2O_2CO_3$ at 23.90°(011), 32.73°(110), 35.31°(112), 39.51°(006), 40.3 6°(015), 42.30°(114), 48.93°(022), 53.41°(121), 54.51°(024), 56.90°(123). This is good for the Fe_2O_3 modification on the surface of $Bi_2O_2CO_3$.Furthermore, the lattice fringe of $Bi_2O_2CO_3$ and Fe_2O_3 match es so well that it could be ascribed to mixture of the characteristic peak of $Bi_2O_2CO_3$ and Fe_2O_3. Howe ver, the new occured impurity peak should be confirmed by the following characterization ways.

3.2. Morphology characterization

It was found that $Bi_2O_2CO_3$ (Figure 2a) and $Fe_2O_3/Bi_2O_2CO_3$ (Figure 2b) showed the similar morphological structure as nano-chips, which could be accounted for the limited doping amount of Fe_2O_3. The SEM further indicated that the catalyst powder was uniformly dispersed without large-scale agglomeration.In addition, the results of TEM indicate that the catalyst has a length of about 250 nm and the thickness is about 20 nm. This is consistent with existing literature reports [12].

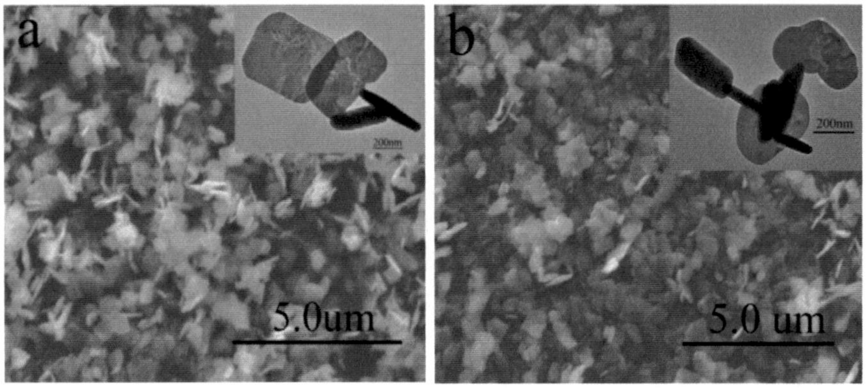

Figure 2. SEM and TEM (inserted graph) of pure $Bi_2O_2CO_3$ (a) and 1.5 $Fe_2O_3/Bi_2O_2CO_3$(b).

3.3. UV-Vis-DRS analysis

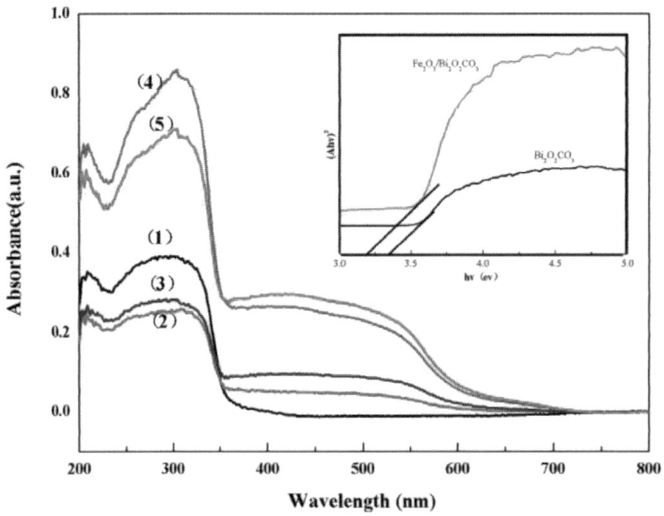

Figure 3. UV-Vis-DRS absorption spectra of $Bi_2O_2CO_3$ (1),0.1 $Fe_2O_3/Bi_2O_2CO_3$ (2),1.0 $Fe_2O_3/Bi_2O_2CO_3$(3),1.5 $Fe_2O_3/Bi_2O_2CO_3$ (4),2.0 $Fe_2O_3/Bi_2O_2CO_3$ (5) and the forbidden band width of $Bi_2O_2CO_3$ and $Fe_2O_3/Bi_2O_2CO_3$.

It is very important to study the UV-vis absorption spectra as the semiconductor band structure is an important factor in determining the photocatalytic activity. Figure 3 shows the absorption spectra of different ratios of Fe_2O_3 in the composite catalyst. As can be seen from the figure, the pure $Bi_2O_2CO_3$ can only absorb UV light (lambda is about 360 nm),and with the increase of Fe_2O_3 deposition, the absorption peak of Fe_2O_3 is seen at lambda equals to 380 nm. The co-existence of Fe_2O_3 and $Bi_2O_2CO_3$ also indicates that the final composite catalyst has Fe_2O_3. In addition,the band width Eg follows the folmula.

$$(\alpha hv)^{1/n}=C(hv-E_g) \tag{1}$$

Where alpha is the absorption coefficient,h is the Planck constant,v is the frequency, C is constant. The index n is directly related to the type of semiconductor, direct band-gap semiconductor n=1/2, indirect band-gap semiconductor n=2 [12]. It is known by the lambert-beer law that A is proportional to alpha, and A is the absorbance in the visible diffuse reflection.Hv(denoted by 1240/wavelength) is the x-coordinate. The tangent line of $(\alpha hv)^{1/n}$ is the graph of the vertical coordinate, and the tangent line of the curve is equal to Eg. Figure 3 shows that the forbidden band width of the composite catalyst after doping was less than that of the pure $Bi_2O_2CO_3$. The electrons are more easily excited to produce the luminous carrier [4].

3.4. Effect of Fe_2O_3 loading on the Fenton-like reactivity of $Bi_2O_2CO_3$

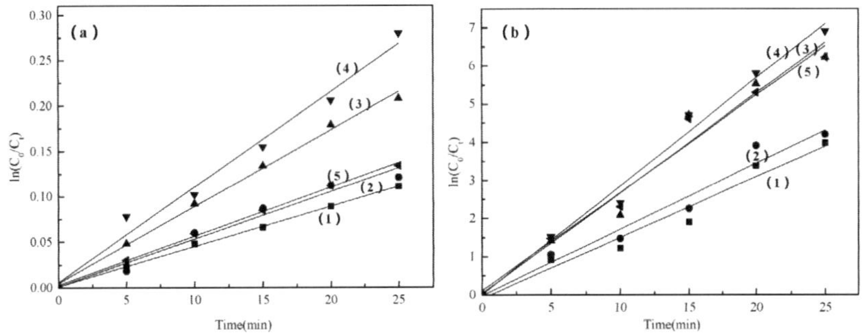

(a)in the dark (b)Under visible light

Figure 4. Fenton-like degradation of RhB with $Bi_2O_2CO_3$(1),0.1 $Fe_2O_3/Bi_2O_2CO_3$(2),1.0 $Fe_2O_3/Bi_2O_2CO_3$(3),1.5 $Fe_2O_3/Bi_2O_2CO_3$ (4),2.0 $Fe_2O_3/Bi_2O_2CO_3$(5).

Table 1. Degradation rate constants of RhB of the $Bi_2O_2CO_3$ and $Fe_2O_3/Bi_2O_2CO_3$ catalysts.

Samples	k/min^{-1}	
	In the dark	Under visible light
$Bi_2O_2CO_3$	0.4	15.9
0.1 $Fe_2O_3/Bi_2O_2CO_3$	0.5	17.4
1.0 $Fe_2O_3/Bi_2O_2CO_3$	0.8	26.3
1.5 $Fe_2O_3/Bi_2O_2CO_3$	1.0	28.4
2.0 $Fe_2O_3/Bi_2O_2CO_3$	0.5	25.6

Figure 4 investigates the degradation effect of RhB with various $Fe_2O_3/Bi_2O_2CO_3$ catalytic systems . The Fenton-like response in the dark condition was achieved in Figure 4a, and the photo-Fenton-like reactivity under visible light irradiation was concluded in Figure 4b.The corresponding degradation rate constants of RhB of the $Bi_2O_2CO_3$ and $Bi_2O_2CO_3$-Fe_2O_3 catalysts are listed in Table 1.

In Figure 4a, the Fenton-like response conforms to the first order dynamic equation.With the increase of Fe_2O_3 loading,the catalytic performance of the catalyst gradually increased, the maximum removal efficiency occured in the system of 1.5 $Fe_2O_3/Bi_2O_2CO_3$.The reaction rate of 1.5 $Fe_2O_3/Bi_2O_2CO_3$ is 2.5 times than that of $Bi_2O_2CO_3$. This indicates that a small amount of Fe_2O_3 doping is beneficial to the enhancement of Fenton-like effect Due to the Fe^{3+} components in the complex $Fe_2O_3/Bi_2O_2CO_3$, the Fenton reaction is bound to occur with the addition of H_2O_2 as a catalytic agent.The RhB is partially oxidized and degraded.

Figure 4b shows the degradation effect of RhB under photo-Fenton-like degradation of pure $Bi_2O_2CO_3$ and $Fe_2O_3/Bi_2O_2CO_3$ It was found that pure $Bi_2O_2CO_3$ shows a good photo-Fenton catalytic effect on RhB. Moreover, the addition of Fe_2O_3 enhances the photo-Fenton catalytic effect of the catalyst.Among them, the catalytic effect of 1.5 $Fe_2O_3/Bi_2O_2CO_3$ acheived the highest treatment efficiency, and the catalytic effect is 1.8 times that of pure $Bi_2O_2CO_3$. Notably,the photo-Fenton-like degradation effect of composite catalyst 1.5 $Fe_2O_3/Bi_2O_2CO_3$ under visible light is 71 times than that of pure $Bi_2O_2CO_3$ under Fenton-like system in the dark.

Because of the response capacity of $Bi_2O_2CO_3$ to visible light, and its microstructure can increase the catalytic activity point, the synergistic effect of these advantages and Fenton reaction make photocatalytic activity and degradation ability growing rapidly.This may be due to the fact that the photoelectric electron generated by Fe_2O_3 is more likely to be transferred from its guide band to the guide band of $Bi_2O_2CO_3$ [12]. The electron in the guide band of $Bi_2O_2CO_3$ can react with the adsorption of O_2 in the dye solution through a series of reactions to generate H_2O_2, the accumulation of holes in the Fe_2O_3 price band can react with the OH^- in the dye solution, then it produces ·OH with strong oxidation which is highly oxidized, while H_2O_2 can also quickly obtained an electron which also produces the strong oxidizing of ·OH [11].

4. Conclusions

In this paper, $Bi_2O_2CO_3$ and $Fe_2O_3/Bi_2O_2CO_3$ composite photocatalyst were successfully prepared by hydrothermal method.The final composite catalysts have a tetragonal phase with a nano-chip structure.The length of the composite catalysts is about 250 nm and the width of the thickness is about 20 nm. The addition of Fe_2O_3 can accelerates the degradation efficiency of RhB dramatically. The results further reveals that $Fe_2O_3/Bi_2O_2CO_3$ acts as a photo and a Fenton-like catalyst as well. It is expected to shed a new idea for the constructing novel composite photocatalyst and also provide a potential method for the removal of dyes in the aqueous system.

Acknowledgements

This work was supported by National Natural Science Foundation of China (Grant no. 21507067).

References

[1] Cong R, Sun J, Yang T, et al. ChemInform Abstract: Syntheses and Crystal Structures of Two New Bismuth Hydroxyl Borates Containing $[Bi_2O_2]^{2+}$ Layers: $Bi_2O_2[B_3O_5(OH)]$ and $Bi_2O_2[BO_2(OH)]$.[J]. Cheminform, 2011, 42(32):no-no.

[2] Yu C, Yu J C, Fan C, et al. Synthesis and characterization of Pt/BiOI nanoplate catalyst with enhanced activity under visible light irradiation[J]. Materials Science & Engineering B, 2010, 166(3):213-219.

[3] Chi C T, Lu I F, Chiu I C, et al. Flexible Transparent ZnO:Al/ZnO/CuAlOx :Ca Heterojunction Diodes on Polyethylene Terephthalate Substrates[J]. Journal of Electronic Materials,2013, 42(6):1242-1245.

[4] Tian N, Huang H, Guo Y, et al. Ag-$C_3N_4/Bi_2O_2CO_3$, composite with high visible-light-driven photocatalytic activity for rhodamine B degradation[J]. Applied Surface Science,2014, 322:249-254.

[5] Yan Z, Fang D, Chen M, et al. Synthetic $Bi_2O_2CO_3$, nanostructures: Novel photocatalyst with controlled special surface exposed[J].Journal of Molecular Catalysis A Chemical,2010,317 (1–2):34-40.

[6] Lin X, Guo X, Shi W, et al. Hydrothermal Synthesis of Mesh-Like $Bi_2O_2CO_3/Bi_2MoO_6$ Heterojunction with Enhanced Photocatalytic Activity[J].Nanoscience&Nanotechnology Letters, 2015, 7(9):691-696.

[7] Zai J, Cao F, Liang N, et al. Rose-like I-doped $Bi_2O_2CO_3$, microspheres with enhanced visible light response: DFT calculation, synthesis and photocatalytic performance[J].Journal of Hazardous Materials, 2016, 321:464.

[8] Lv X, Tang Z L, Zhai X L, et al. Preparation and photocatalytic activities of $Bi_2O_2CO_3/TiO_2$ composite powders[J]. Gongneng Cailiao/journal of Functional Materials, 2013, 44(12):1812-1815.

[9] Wei Z, Wei X, Wang S, et al. Preparation and visible-light photocatalytic activity of α-Fe_2O_3/γ-Fe_2O_3, magnetic heterophase photocatalyst[J]. Materials Letters, 2014, 118:107-110.

[10] Goi A, Trapido M.Hydrogen peroxide photolysis, Fenton reagent and photo-Fenton for the degradation of nitrophenols: a comparative study[J]. Chemosphere, 2002, 46(6):913-22.

[11] Nogueira R F,Oliveira M C, Paterlini W C.Simple and fast spectrophotometric determination of H_2O_2 in photo-Fenton reactions using metavanadate.[J]. Talanta,2005,66(1):86.

[12] Hu D, Zhang K, Yang Q, et al. Super-high photocatalytic activity of Fe_2O_3, nanoparticles anchored on $Bi_2O_2CO_3$, nanosheets with exposed {001} active facets[J]. Applied Surface Science, 2014, 316(1):93-101.

Construction of High Activity Titanium Dioxide Crystal Surface Heterostructures and Characterization of Its Basic Properties

Chunxiao Wang[1], DanQi Li[1], Tingting Shen[1], Cheng Lu[1], Jing Sun[1]* and Xikui Wang[2]*

[1]Department of Environmental Science and Engineering, Qilu University of Technology, Jinan 250353, China

[2]Chemistry and Pharmaceutical Engineering, Qilu University of Technology, Jinan 250353, China

*xk_wang@qlu.edu.cn

Abstract Heterogeneous photocatalytic materials, which combine the advantages of photocatalytic materials and heterojunction, have been developed rapidly in the field of environmental pollution control. In this paper, TiO_2 surface heterojunction catalysts with different catalytic activity were prepared by controlling the amount of HF, and their XRD characterization was also carried out. In addition, the optimum amount of HF was determined by photocatalytic degradation of simulated dye wastewater by methylene blue solution. And the optimal amount of catalyst and the optimal pH reaction conditions for degradation experiments were used to screen the highly reactive titania surface heterojunction system and its optimum application conditions. It provides the possibility of application in the degradation of industrial wastewater and environmental treatment.

1. Introduction

TiO_2 is a wide bandgap n-type semiconductor, with high energy absorption of ultraviolet light has a strong absorption capacity.When TiO_2 is irradiated with light, an electron-hole pair is formed [1], and OH^- and H_2O adsorbed on the surface of TiO_2 are oxidized to produce active substances such as hydroxy radical (·OH) having strong oxidizing properties, and OH is added to the organic matter in water Mineralized into CO_2 and H_2O [2, 3].TiO_2 is a non-toxic, stable performance of the new catalyst, it is no degradation of organic compounds in water, no secondary pollution. Nano-TiO_2 has a huge surface area, the greatest degree of degradation of organic matter in wastewater, widely used in industrial wastewater treatment and new energy development and utilization [4].

However, TiO_2 as a wide bandgap material, will only be short wavelength of ultraviolet light excitation, which is the use of ultraviolet light content of only 4% of the sun is extremely unfavorable [5, 6]. In addition, the photo-generated electron-hole generated by the use of light is high, so that the photonics efficiency is low and the photocatalytic performance is not prominent [7, 8]. This is also the main reason for the limitation of the photocatalyst of the semiconductor type in the practical application. While the heterogeneous junction of the built-in electric field can inhibit the photo-charge charge, improve the quantum efficiency, if the TiO_2 and narrow-band semiconductors constitute heterojunction, narrow-band semiconductor sensitization can expand the response spectrum of TiO_2, promising to overcome the above-mentioned TiO_2 Disadvantages [9, 10].

The interfacial regions formed by the contact of two different semiconductors are heterojunction, and the photocatalytic properties of TiO_2 with different crystal phases are different, which is due to the different structure [11, 12]. Nano-heterojunction photocatalytic materials combine the advantages of nanomaterials and heterojunctions and are rapidly developing in the field of environmental pollution control [13, 14]. Zheng et al. Successfully synthesized the titanium dioxide crystals with the {001} ratio of 45%, 69% and 82% by hydrothermal method, and the yield of active hydroxyl groups (·OH) and the yield of H_2 were characterized by three Photocatalytic properties of materials. The results show that the higher the proportion of {001} crystal face, the lower the photocatalytic performance of the material [15, 16].

In this paper, we studied the preparation of TiO_2 surface heterojunction with different photocatalytic properties by controlling the addition of HF, and explored the influence factors of catalyst dosage and pH during the photocatalytic degradation of methylene blue solution, and screened the best TiO_2 crystal face Heterojunction system and its optimal reaction conditions.

2. Experimental

Titanic acid four butyl ester ($C_{16}H_{36}O_4Ti$, chemical purity, Sinopharm Chemical Reagent Co., Ltd.). Hydrofluoric acid (HF, analytical pure, Aladdin Industrial Corporation). Methylene blue ($C_{16}H_{18}C_lN_3S$, $3H_2O$, analysis of pure, Tianjin Municipality kemi'ou Chemical Reagent Co., Ltd.). Anhydrous ethanol (C_2H_5OH, analytical purity, Tianjin Fuyu Fine Chemical Co., Ltd.). The water used in the experiment was deionized water.

2.1 Preparation of Titanium Dioxide

2.5 ml, 3 ml, 3.5 ml and 4 ml hydrofluoric acid were added into the plastic beaker containing 25 ml titanic acid four butyl ester, and the beaker was placed on a magnetic stirrer to stir 30 min to make it dissolve evenly. Then, the solution in the beaker was transferred into the reaction kettle and reacted in 180℃ for 24h to obtain white precipitate. The white precipitate was washed three times with anhydrous ethanol and deionized water. The impurities were removed, and then the TiO_2 precipitate was obtained. Four different titanium dioxide crystal powders were obtained by freeze drying of TiO_2.

2.2 Characterization

The samples structure was characterized by powder X-ray diffraction (XRD) using a scanning rate of 0.05 °/s within a 2θ range of 10°–80° (Bruker D8 Advance; Cu Kα=1.5404 Å) [17].

2.3 Determination of photocatalytic activity

The photocatalytic activity was determined by absorbance changes of methylene blue solution (MB solution, 10 ppm) in a certain amount of photocatalytic reactor [18]. 500 W xenon lamp with internal irradiation was used as light source. And 250 mL beaker as the reactor loading reaction solution. The distance between the center of the light source and the beaker mouth is about 5 cm. Before the photocatalytic experiment, the mixed solution was stirred for 30 minutes in the absence of light so as to make the mixed solution reach the adsorption equilibrium. During the experiment, the magnetic stirrer at the bottom of the beaker continues to stir the reaction solution to ensure that the reaction solution and the catalyst are mixed evenly during the experiment. 0.125 g photocatalyst was added into 150 mL MB solution to form suspension and homogenized by ultrasonic vibration. 5 mL mixed solution was taken at intervals of 10 minutes and filtered by 0.45 μm microporous filter. Absorbance was measured at 664 nm by visible spectrophotometer.

3. Results and discussion

3.1 Characterization of the TiO_2 Crystal Surface Heterostructures

The crystal shape and crystallinity of the prepared samples were further analyzed by XRD. As shown in Fig. 1, the XRD pattern of the prepared titanium dioxide heterojunction was obtained when the addition amount of hydrofluoric acid was 2.5 mL, 3 mL, 3.5 mL, 4 mL. It can be concluded from the figure, the diffraction peak of TiO_2 at 25.3°, 38.1°, 48°, 53.9°, 54.8°, 62.8°, 69.4°, 70.5° and 75.1° respectively corresponding to the crystal phase (101), (004), (200), (105), (211), (204), (116), (220) and (215) crystal [19]. All the peaks are in accordance with the standard card (JCPDS card number: 73-1764). Four kinds of TiO_2 crystalline surface heterostructures are pure anatase phase of TiO_2. The Miscellaneous peak of TiO_2 crystal heterojunction with HF addition of 3.5 mL and 4 mL is caused by excessive concentration of F^-.

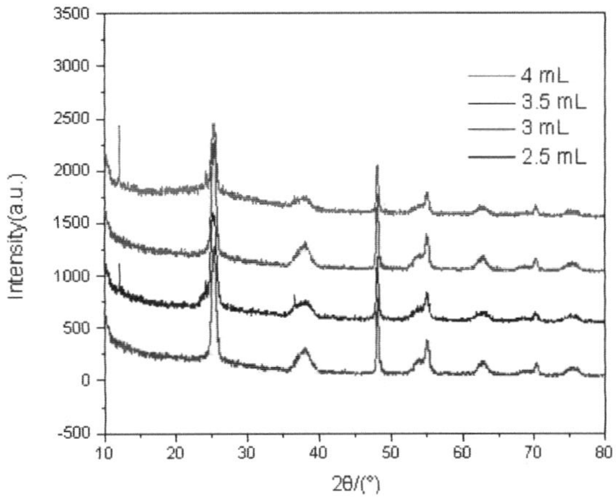

Figure 1. XRD Characterization of Titanium Dioxide under Different Preparation Conditions. 4 mL:Hydrofluoric acid was added in an amount of 4 mL; 3 mL:Hydrofluoric acid was added in an amount of 3 mL; 3.5 mL:Hydrofluoric acid was added in an amount of 3.5 mL; 2.5 mL:Hydrofluoric acid was added in an amount of 2.5 mL;

Normally, the crystal structure of anatase TiO_2 is a cross section eight body double cone, which includes eight {101} surfaces on the side and two {001} surfaces at the top and bottom. Research shows that with the increase of HF content, the intensity of XRD peak and XRD peak samples is more stable, the width is more and more narrow, indicating that TiO_2 crystallinity increases, the percentage of {001} is more and more high that F- plays a key role in the formation of TiO_2 Nano plate. [20]. The {001} percentage of the corresponding amount of HF is shown in Table 1.

Table 1. Physicochemical Properties of Different Samples

sample	HF(mL)	{001}facets%
HF2.5	2.5	46
HF3	3	49
HF3.5	3.5	52
HF4	4	55

3.2 Photocatalytic activity

The photocatalytic activity of the catalyst was determined by the degradation of methylene blue solution.

Figure. 2(a). Results of photocatalytic degradation of methylene blue solution. by TiO_2 crystal surface heterojunction prepared with different HF addition. It can be seen from the diagram that when the amount of HF increases gradually from 0-3 mL, the photocatalytic activity also increases; When the amount of HF is more than 3 mL, with the increase of HF, the photocatalytic activity decreases gradually, This is because the appropriate ratio of the {001} and the {101} surface can significantly reduce the recombination rate of the photoinduced electron and hole. However, the HF value greater than 3 leads to a higher percentage of {001} surface, which makes recombination recombination of electron holes. Therefore, the best amount of HF added to the photocatalytic activity of TiO_2 crystal surface heterojunction is 3mL.

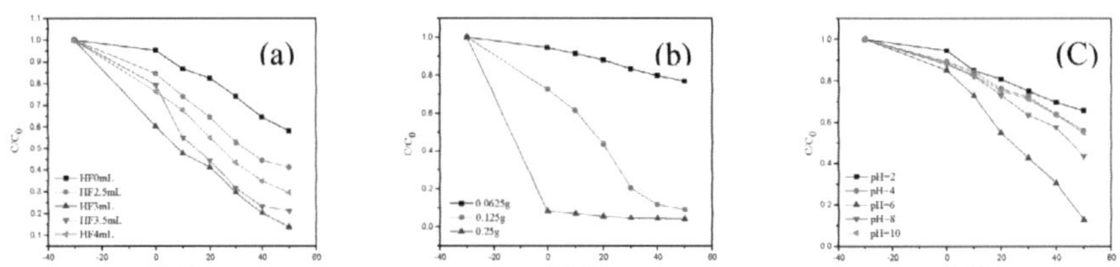

Figure.2(a) Photocatalytic Performance of different TiO_2 catalysts; **(b)** catalytic effect diagram of different dosages of catalyst; **(c)** Influence Diagrams of different pH on catalytic performance of catalysts

Figure. 2(b) is the result of the influence of the dosage of TiO_2 catalyst on the photocatalytic performance. According to the diagram, the degradation effect of methylene blue solution is not significant when the dosage of catalyst is 0.0625 g. In addition, when 0.25 g catalyst was used to degrade methylene blue solution, the whole degradation process was almost completed in the adsorption stage of the first 30 min. Therefore, the experimental results have no reference value. When the dosage of catalyst is 0.125 g, the concentration of methylene blue solution decreases gradually with the increase of time, and the effect is outstanding, so 0.125 g is the best catalyst dosage.

Figure 2(c) is a comparison of photocatalytic performance of TiO_2 catalyst under different pH conditions. It can be seen from the diagram that when the methylene blue solution pH=2 gradually increases to pH=6, the photocatalytic activity also increases. When the pH value is greater than 6, the photocatalytic activity decreases with the increase of pH value. Therefore, the optimum pH value of TiO_2 catalyst for methylene blue solution degradation was 6.

4. Conclusions

In this paper, TiO_2 crystal surface heterojunction catalyst with different catalytic activity was prepared by controlling the amount of HF. The optimum amount of HF was determined by photocatalytic degradation of methylene blue solution. And the optimal dosage of catalyst and the optimal pH reaction conditions for degradation experiments. The final results are as follows：

(1) When the amount of HF is 3 mL, the photocatalytic performance of the TiO_2 crystal surface heterojunction catalyst is the best.

(2) The optimum dosage of catalyst for photocatalytic degradation of methylene blue solution was 1.25 g.

(3) Optimum pH=6 for photocatalytic degradation of methylene blue solution.

Acknowledgment

This work was supported by the National Nature Science Foundation of China (No. 21507067) and Open Fund of the Key Laboratory of Pulp and Paper Science & Technology of the Ministry of Education in Qilu University of Technology (No. KF201416).

References

[1 Tamaki, Y., et al., *Dynamics of efficient electron-hole separation in TiO_2 nanoparticles revealed by femtosecond transient absorption spectroscopy under the weak-excitation condition.* Physical Chemistry Chemical Physics Pccp, 2007. **9**(12): p. 1453.

[2] Liu, B.S., et al., *[The surface states and the electron-hole pair recombination of TiO_2 nanopowders].* Spectroscopy & Spectral Analysis, 2006. **26**(2): p. 208.

[3] Bu, Y. and Z. Chen, *Effect of oxygen-doped C3N4 on the separation capability of the photoinduced electron-hole pairs generated by O-C3N4@TiO2 with quasi-shell-core nanostructure.* Electrochimica Acta, 2014. **144**(144): p. 42-49.

[4] Fujishima, A., *TiO_2 Photocatalysis and Related Surface Phenomena.* Surface Science Reports, 2008. **63**(12): p. 515-582.

[5] Fujishima, A. *TiO_2 Photocatalysis and Related Surface Phenomena.* in *the 60th annual meeting of the international society of electrochemistry.* 2009.

[6] Behnajady, M.A., et al., *The effect of particle size and crystal structure of titanium dioxide nanoparticles on the photocatalytic properties.* J Environ Sci Health A Tox Hazard Subst Environ Eng, 2008. **43**(5): p. 460-467.

[7] Yang, H.G., et al., *Anatase TiO_2 single crystals with a large percentage of reactive facets.* Nature, 2008. **453**(7195): p. 638.

[8] Villalobos-Hernández, J.R. and C.C. Müller-Goymann, *Sun protection enhancement of titanium dioxide crystals by the use of carnauba wax nanoparticles: the synergistic interaction between organic and inorganic sunscreens at nanoscale.* International Journal of Pharmaceutics, 2006. **322**(1–2): p. 161-170.

[9] Linsebigler, A.L., G. Lu, and J.T. Yates, *Photocatalysis on TiO_2 Surfaces: Principles, Mechanisms, and Selected Results.* Chemical Reviews, 1995. **95**(3): p. 735-758.

[10] Fahmi, A., et al., *Theoretical analysis of the structures of titanium dioxide crystals.* Physical Review B Condensed Matter, 1993. **47**(18): p. 11717.

[11] Tachikawa, T., S. Yamashita, and T. Majima, *Evidence for crystal-face-dependent TiO_2 photocatalysis from single-molecule imaging and kinetic analysis.* Journal of the American Chemical Society, 2011. **133**(18): p. 7197.

[12] Akihito Imanishi, §, et al., *Crystal-Face Dependence and Photoetching-Induced Increases of Dye-Sensitized Photocurrents at Single-Crystal Rutile TiO_2 Surfaces.* Journal of Physical Chemistry B, 2006. **110**(42): p. 21050.

[13] Zhang, J., et al., *Novel dual heterojunction between MoS2 and anatase TiO_2 with coexposed {101} and {001} facets.* Journal of the American Ceramic Society, 2017.

[14] Choi, D.J., et al., *Dye Sensitization of Four Low Index TiO_2 Single Crystal Photoelectrodes with a Series of Dicarboxylated Cyanine Dyes.* Langmuir the Acs Journal of Surfaces & Colloids, 2013. **29**(30): p. 9410-9.

[15] Paulose, M., et al., *Application of highly-ordered TiO_2 nanotube-arrays in heterojunction dye-sensitized solar cells.* Journal of Physics D Applied Physics, 2006. **39**(12): p. 2498.

[16] Yang, L., et al., *High Efficient Photocatalytic Degradation of p-Nitrophenol on a Unique Cu2O/TiO_2 p-n Heterojunction Network Catalyst.* Environmental Science & Technology, 2010. **44**(19): p. 7641-6.

[17] Gao, X.F., et al., *An Efficient Method To Form Heterojunction CdS/TiO_2 Photoelectrodes Using Highly Ordered TiO_2 Nanotube Array Films.* Journal of Physical Chemistry C, 2009. **113**(47): p. 20481-20485.

[18] Shen, S., et al., *Effect of Noble Metal in CdS/M/TiO_2 for Photocatalytic Degradation of Methylene Blue under Visible Light.* International Journal of Green Nanotechnology Materials Science & Engineering, 2010. **1**(2): p. M94-M104.

[19] Watanabe, S., X. Ma, and C. Song, *Characterization of Structural and Surface Properties of Nanocrystalline TiO_2−CeO2 Mixed Oxides by XRD, XPS, TPR, and TPD.* Journal of Physical Chemistry C, 2009. **113**(32): p. 14249-14257.

[20] Yu, J., et al., *Enhanced photocatalytic CO_2-reduction activity of anatase TiO_2 by coexposed {001} and {101} facets.* Journal of the American Chemical Society, 2014. **136**(25): p. 8839-42.

IOP Conf. Series: Earth and Environmental Science **112** (2018) 012011 doi:10.1088/1755-1315/112/1/012011

An Energy Integrated Dispatching Strategy of Multi- energy Based on Energy Internet

Weixia Jin[1,*] and Jun Han[1]

[1]Tianjin University of Technology, No. 391 Binshui Road, Xiqing District, Tianjin, China.

*zhilixiang111@163.com

Abstract. Energy internet is a new way of energy use. Energy internet achieves energy efficiency and low cost by scheduling a variety of different forms of energy. Particle Swarm Optimization (PSO) is an advanced algorithm with few parameters, high computational precision and fast convergence speed. By improving the parameters ω, c1 and c2, PSO can improve the convergence speed and calculation accuracy. The objective of optimizing model is lowest cost of fuel, which can meet the load of electricity, heat and cold after all the renewable energy is received. Due to the different energy structure and price in different regions, the optimization strategy needs to be determined according to the algorithm and model.

1. Introduction

At present, energy that used by people are mainly fossil fuels, such as coal, oil, and natural gas. The large use of fossil fuels raises serious problems, such as energy shortages and environmental pollution. And these problems have a negative impact on people's lives [1]. With the development of human society, energy and environmental issues are getting more and more attention. In order to solve this problem, we should start from two aspects. On the one hand, it is necessary to improve the efficiency of fossil energy and reduce fossil energy consumption. On the other hand, we must continue to develop the use of clean energy [2]. The foundation of energy internet is renewable energy which can achieve a comprehensive utilization of various forms of energy including electricity, heating and cooling. The purpose of energy interconnection is to improve the overall efficiency and reduce the cost. Energy router as energy dispatch equipment of energy internet needs to consider different models of energy forms and then use advanced algorithms to achieve efficient use of energy [3].

2. Five Models of Energy Involved in Energy Internet

The hot and cold power supply system can use different forms of energy to generate electricity, heat and cold. The electrical load of the system is provided by the grid and the power generation equipment [4].

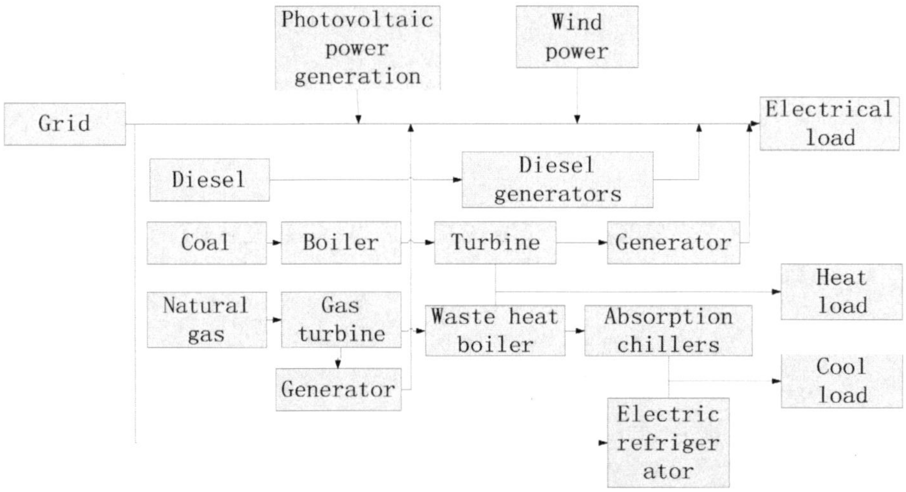

Figure 1. The structure of energy dispatch

The power generation equipment includes photovoltaic, wind power, coal, natural gas and diesel emergency power generation. The heat load is provided by the extraction of the steam turbine in the coal and the exhaust heat from the gas. The cold load is provided by an absorption chiller in the gas power generation, and electric chiller is equipped when the cold load is big [5]. The structure of energy dispatch is showed in Fig 1.

2.1 Model of Wind Power

$$P_{wind} = \begin{cases} 0, & v < v_{in}, v \geq v_{out} \\ P_{wind0} \times \frac{v - v_{in}}{v_0 - v_{in}}, & v_{in} \leq v \leq v_0 \\ P_0, & v_0 < v < v_{out} \end{cases} \quad (1)$$

In the above formula, P_{wind} is the power of the wind turbine, P_{wind0} is the rated power of the wind turbine, v is the actual wind speed, v_0 is the rated wind speed, v_{in} is the cut in wind speed, and v_{out} is the cut off wind speed.

Wind power is directly affected by the wind speed, while the speed of the natural wind has a lot of randomness and volatility [6]. In the energy Internet, wind power as clean energy should be accepted by the grid completely. So wind power constraints should be set:

$$P_{wind} = P_{wind}^{for} \quad (2)$$

$$0 \leq P_{wind} \leq P_{wind0} \quad (3)$$

Among them, P_{wind}^{for} is used to indicate the predicted power of the wind power, that is to say, the wind power is involved in the dispatching according to the predicted power, and the electricity is accepted by the grid completely.

2.2 Model of Photovoltaic Power

For facilitate calculation, a simple model is used in the optimization, that is, the output power is only affected by light and temperature.

$$P_{solar} = \frac{P_{solar0} \times G \times [1 + k_T(T - T_0)]}{G_0} \quad (4)$$

P_{solar} is the power of the photovoltaic panels, P_{solar0} is the maximum power of the photovoltaic panels in the test environment(25 ℃, 1000W / m^2), G is the real-time light intensity, k_T is the temperature coefficient, T is the real-time temperature, T_0 is Photovoltaic board design reference temperature, G_0 is standard light intensity of test.

As wind power, photovoltaic power generation is affected by the environment, so photovoltaic power generation also has great uncertainty and volatility [7]. In the energy internet, photovoltaic power generation as clean energy, its power needs to be accepted by grid completely.

$$P_{solar} = P_{solar}^{for} \quad (5)$$

$$0 \leq P_{solar} \leq P_{solar0} \quad (6)$$

P_{solar}^{for} is the predicted power of photovoltaic panel.

2.3 *Model of Cogeneration*

The relationship between coal consumption and electricity load is quadratic function in cogeneration, and that is a kind of highly efficient coal-fired power generation technology [8]. In order to make as much use of coal energy, cogeneration units decide the load of electricity by heat.

The amount of coal consumed by the coal-fired unit:

$$F_{coal} = a_{coalf} + b_{coalf}P_{coal} + c_{coalf}P_{coal}^2 \quad (7)$$

F_{coal} is the amount of coal consumed by the unit, P_{coal} is the electrical power of the unit, a_{coalf}, b_{coalf}, c_{coalf} is the coefficient of the system.

Heat supply of coal - fired units:

$$Q_{coal} = P_{coal} \times \eta_{qcoal} \times \varphi_{qcoal} \times \theta_{qcoal} \quad (8)$$

Q_{coal} is the heating rate per unit time of the system, η_{qcoal} is the heating coefficient corresponding to the electric load, φ_{qcoal} is the ratio of the heat load required for the current time system to the maximum heat that can be provided, θ_{qcoal} is the heat conversion efficiency.

Thermal power generation is a very mature way of generating electricity, compared to new energy power generation, thermal power is controllable. The constraints are mainly reflected in the electrical power and thermal power, that is, there is a minimum power limit.

$$\varphi_{cut,coal}P_{coal0} \leq P_{coal} \leq P_{coal0} \quad (9)$$

$\Phi_{cut,\,coal}$ is the unit's minimum power factor, P_{coal0} is the rated power of unit.

$$0 \leq Q_{coal} \leq Q_{coal0} \quad (10)$$

Q_{coal0} is the maximum heat supply in per time.

2.4 *Model of Gas Power Generation System*

The gas power system can provide electricity, heat and cooling, and its energy efficiency is higher. The efficiency of the system decreases with the decrease of the output. At present, third-order efficiency model is used to analyze the efficiency of the gas turbine [9].

$$F_{gas} = a_{gasf} + b_{gasf}P_{gas} + c_{gasf}P_{gas}^2 + d_{gasf}P_{gas}^3 \quad (11)$$

F_{gas} is the amount of natural gas consumed per time, P_{gas} is the electricity power of the gas turbine, a_{gasf}, b_{gasf}, c_{gasf}, d_{gasf} is the coefficient of the system.

The heating of the gas unit is mainly provided by the exhaust heat regenerator, whose power is affected by the power of heating load and the output of the gas turbine.

$$Q_{gas} = P_{gas} \times \eta_{qgas} \quad (12)$$

$$Q_{qgas} = Q_{gas} \times \varphi_{qgas} \times \theta_{qgas} \quad (13)$$

Q_{gas} is the heat of exhaust, η_{qgas} is the relationship between the exhaust heat and the electric load, Q_{qgas} is the heating load provided by the gas turbine, φ_{qgas} is the proportion of the heat load provided by the gas turbine, θ_{qgas} is conversion efficiency of heating.

The cooling of the gas unit is mainly provided by an absorption chiller, and the absorption chiller needs to be cooled by the heat in the exhaust gas.

$$Q_{cgas} = Q_{gas} \times \varphi_{cgas} \times \theta_{cgas} \quad (14)$$

Q_{cgas} is the amount of cooling provided per time of the gas turbine unit, φ_{cgas} is the ratio of the total heat of the exhaust gas used for the heat of cooling, and θ_{cgas} is the conversion efficiency between heat and cooling.

As with coal-fired units, the gas unit has the smallest output limit, that is, the power load must be greater than a certain value.

$$\varphi_{cut,gas}P_{gas} \leq P_{gas} \leq P_{gas0} \tag{15}$$

$\Phi_{cut,\ gas}$ is the minimum power coefficient of unit, P_{gas0} is the rated power of unit.

Heating load constraints:

$$0 \leq Q_{qgas} \leq Q_{qgas0} \tag{16}$$

Q_{qgas0} is the rated calorific value of the load.

Cooling load constraint:

$$0 \leq Q_{cgas} \leq Q_{cgas0} \tag{17}$$

Q_{cgas0} is the rated cooling capacity of the load.

$$\varphi_{qgas}+\varphi_{cgas} \leq 1 \tag{18}$$

The heat used for heating and cooling cannot exceed the total heat in the exhaust.

2.5 Model of Oil Power Generation System

Oil power generation system is generally used as an emergency power supply in a country that lacks oil, and the relationship between fuel consumption and power is quadratic.

$$F_{oil} = a_{oilf} + b_{oilf}P_{oil} + c_{oilf}P_{oil}^2 \tag{19}$$

F_{oil} is the consumption of unit fuel, P_{oil} is the unit's electrical power, a_{oilf}, b_{oilf}, c_{oilf} is the coefficient of system.

Constraints of fuel power system:

$$\varphi_{cut,oil}P_{oil0} \leq P_{oil} \leq P_{oil0} \tag{20}$$

$\Phi_{cut,\ oil}$ is the coefficient of minimum power, P_{oil0} is the rated power.

2.6 Constraints of System

When the power system is in operation, it is necessary to keep the amount of electricity equal to the load in real time:

$$P_{wind} + P_{solar} + P_{coal} + P_{gas} + P_{oil} = P_{need} \tag{21}$$

P_{need} is the power load of system.

The heating system needs to maintain a balance of heat load; otherwise it will cause the temperature rise or fall:

$$Q_{qcoal} + Q_{qgas} = Q_{qneed} \tag{22}$$

Q_{qneed} is the heating load of system.

In the actual operation of the system, the cooling load needs to maintain balance. If the cooling provided by gas unit cannot meet the demand, it is necessary to put in electric cooling:

$$Q_{ce} = P_{ce} \times \theta_{ce} \tag{23}$$

Q_{ce} is the cooling capacity for electricity cooling, P_{ce} is the power used for electrical cooling, and θ_{ce} is the conversion efficiency of the electricity to cold.

$$Q_{cgas} + Q_{ce} = Q_{cneed} \tag{24}$$

Q_{cneed} is the system's cooling load.

There are many parameters that affecting the energy internet evaluation indicators, but the key is economic [10]. The original intention of energy internet is to improve energy efficiency and maximize the use of energy. The minimum cost of running is chosen as the optimization target, and the cost includes electricity costs and fuel costs. Set the objective function as:

$$\min C_{total} = C_e + C_{fuel} \tag{25}$$

$$C_e = r_e \times P_e \times \Delta_t \tag{26}$$

$$C_{fuel} = C_{coal} + C_{gas} + C_{oil} \tag{27}$$
$$C_{coal} = r_{coal} \times F_{coal} \times \Delta_t \tag{28}$$
$$F_{coal} = a_{coal} + b_{coal}P_{coal} + c_{coal}P_{coal}^2 \tag{29}$$
$$C_{gas} = r_{gas} \times F_{gas} \times \Delta_t \tag{30}$$
$$F_{gas} = a_{gas} + b_{gas}P_{gas} + c_{gas}P_{gas}^2 + d_{gas}P_{gas}^3 \tag{31}$$
$$C_{oil} = r_{oil} \times F_{oil} \times \Delta_t \tag{32}$$
$$F_{oil} = a_{oil} + b_{oil}P_{oil} + c_{oil}P_{oil}^2 \tag{33}$$

C_{total} represents the total cost of the system, C_e represents the cost of electricity purchased from the grid, r_e represents the electricity price, C_{fuel} represents the total cost of purchasing the fuel, C_{coal} represents the cost of coal purchase, r_{coal} represents the coal price, C_{gas} represents the natural gas purchase cost, r_{gas} represents natural gas price, C_{oil} means oil cost, r_oil means oil price. a_{coal}, b_{coal}, c_{coal}, a_{gas}, b_{gas}, c_{gas}, d_{gas}, a_{oil}, b_{oil}, c_{oil} are the coefficients of the system.

3. Improved Particle Swarm Optimization

PSO simulates the behavior of birds' predation, like a group of birds looking for food; they only know the distance between themselves and food. In order to get food, they will learn from the nearest birds, and gradually close to food.

In order to make the particle swarm algorithm faster convergence, and higher accuracy, improvements could be done as follows:

$$\omega = \omega_{max} - \frac{\omega_{max} - \omega_{min}}{N} \times n \tag{34}$$

ω_{max} and ω_{min} represent the maximum and minimum values of ω, n and N denote the current iteration number and the maximum number of iterations respectively.

$$c_1 = c_{1s} + \frac{n(c_{1e} - c_{1s})}{N} \tag{35}$$
$$c_2 = c_{2e} + \frac{n(c_{2e} - c_{2s})}{N} \tag{36}$$

N and n represent the number of iterations and the maximum number of iterations, c_{1e} and c_{2e} represent the final values of c_1 and c_2, c_{1s} and c_{2s} represent the initial values of c_1 and c_2.

4. Conclusion

For energy internet is based on renewable energy, and energy dispatch needs to ensure renewable energy is accepted. The energy structure and price are different in different regions. So the energy optimization strategy needs to be calculated based on the energy models and algorithms. Through the comprehensive optimization of five forms of energy, it can improve energy efficiency and reduce costs. In the use of Coal-fired units and gas units' waste heat, will not improve their own energy production costs, but can produce benefits by fully used.

Reference

[1] RIFKIN J. The third industrial revolution: how lateral power istransforming energy, *the economy, and the world.* New York: Palgrave MacMillan, 2011.

[2] Junwei Cao and Mingbo Yang. Energy Internet – Towards Smart Grid 2.0. Networking and Distributed Computing (ICNDC), *2013 Fourth International Conference* , 105 – 110.

[3] Yabing Zha, Tao Zhang, Shuren Tan, Zhou Huang and Wenguang Wang. Understanding and thinking on energy internet. *Defense Technology,* 2012,05: 1-6.

[4] Xiong Yan, Wu Jiekang, Wang Qiang, Mao Xiaoming. An Optimization Coordination Model and Solution for Combined Cooling, Heating and Electric Power Systems With Complimentary Generation of Wind, PV, Gas and Energy Storage. Proceedings of the CSEE,2015,14:3616-3625.

[5] Stavridis A, Sinanovic S, Renzo M D, et al. Energy Evaluation of Spatial Modulation at a Multi-Antenna Base Station[C]. Vehicular Technology Conference. IEEE, 2013:1-5.
[6] Henry D, Warren L. Wind Power in Power Systems[J]. IEEE Power Engineering Review, 2014, 22(12):23-27.
[7] Kanchev H, Lu D, Colas F, et al. Energy Management and Operational Planning of a Microgrid With a PV-Based Active Generator for Smart Grid Applications[J]. IEEE Transactions on Industrial Electronics, 2011, 58(10):4583-4592.
[8] Mancarella P, Chicco G. Global and local emission impact assessment of distributed cogeneration systems with partial-load models[J]. Applied Energy, 2009, 86(10):2096-2106.
[9] Kong X Q, Wang R Z, Huang X H. Energy optimization model for a CCHP system with available gas turbines[J]. Applied Thermal Engineering, 2005, 25(2–3):377-391.
[10] A. Huang, M. Crow, G. Heydt, J. Zheng, and S. Dale, "The Future Re- newable Electric Energy Delivery and Management (FREEDM) System:*The Energy Internet,*" RFC 2477, Oct 2011.

Do You Live Happily? Exploring the Impact of Physical Environment on Residents' Sense of Happiness

Chiu-lin Chen[1] and Heng Zhang[1*]

[1]Department of Architecture, National Cheng Kung University, No. 1, University Rd., Tainan City 701, Taiwan

[*]changlin@mail.ncku.edu.tw

Abstract. From the beginning of human civilization, human beings have been exploring the source of happiness. Capitalists believed that economic growth would bring us happiness, but a small, poor Buddhist country—Bhutan—has dispelled the myth with the concept of Gross National Happiness. Today, the concept of happiness has evolved from the field of psychology to all disciplines. Many researchers are exploring how environmental psychology/behavioral psychology influences residents' sense of happiness. The goal of this study is to explore how various environmental factors influence residents' sense of happiness. Through statistical analysis of 473 samples, the results showed that factors like green area, community layout, aesthetics, transportation service and social service significantly influenced residents' sense of happiness. Among all the factors, transportation service and social service simultaneously influence "satisfied" and "joyful", the two factors related to residents' sense of happiness; community layout and aesthetics are significantly correlated to "satisfied"; green area and business service significantly influences "joyful". The results of this study can be used as reference for professionals as they make relevant decisions.

1. Introduction

Aristotle pointed out that "Happiness is the meaning and the purpose of life". However, in the modern society with capitalism as the mainstream, people tend to believe that economic growth will bring us happiness. Gross Domestic Product (GDP) is commonly used in the international community as an important indicator to measure social progress and people's well-being [1]. This trend changed in 2005 when reports showed that 97% of the people in Bhutan, a small and poor Buddhist country, feel happy, despite the fact that they do not own a lot of material enjoyment. King Jigme Dorji Wangchuck of Bhutan proposed the idea of Gross National Happiness (GNH) in 1972. He indicated the four columns of GNH: sustainable and equitable socio-economic development, conservation of environment, preservation and promotion of culture, and good governance [2].

Similarly, the OECD (Organization for Economic Cooperation and Development), with 34 countries which rely on market economy, published "Your Better Life Index" in 2011 and selected material conditions (such as housing, income and wealth, jobs and earnings) as well as quality of life (such as health status, work-life balance, education and skills, social connections, civic engagement and governance, environmental quality, life satisfaction, personal security, subjective well-being) as indicators of people's well-being [3]. Public and private sectors in many countries followed the steps of OCED to investigate measurement of happiness and social progress.

Following the world trends, studies of happiness has shifted from pure psychology to environmental psychology/praxeology, environment and spatial governance. Even planners and designers are studying issues related to happiness, including environmental study, environmental quality, housing, landscape, rural regeneration, leisure and recreation…etc. [4-9]. Over the past few years, one can see abundant papers on the relationship between community environment and residents' sense of happiness. This study explores how physical environmental factors in the community influence residents' sense of happiness. The results of this study can be used as reference for professionals as they make decision related to housing policies, urban planning and design, architecture, landscape, community management.

2. Methodology

2.1. Research site

This research used five communities in southern Taiwan as samples, including Daan New Village, Yiping Neighborhood Community, Qigu Longshan Community, Hongyi Community and Hongnan Community. The first three are located in Tainan City and the others are located in Kaohsiung City. These communities reflect commonly seen communities in southern Taiwan in terms of the number of residents, level of development, community layout and age of houses. The pictures of five communities are shown in figure 1.

Figure 1. The pictures of five communities

2.2. Measures and Procedure

The questionnaires were distributed to survey the relationship between individual participant's sense of happiness and the physical environment of their community. The questionnaires include 10 questions to measure residents' sense of happiness [10], including questions regarding optimism, social control, positive affect, sense of control, physical fitness, satisfaction with self, mental alertness, and working achievements. In the questionnaires, 48 questions are about physical environmental factors which consist of 11 factors: "green areas", "community layout", "architectural scale", "aesthetics", "open space", "street landscape", "transportation service", "business service", "learning service", "social service" and "recreational service". The last part of the questionnaire asks participants questions about their demographic characteristics. This study uses Likert-type scale. In the 5-point scales, 1 represents "strongly disagree", 2 "disagree", 3 "neutral", 4 "agree" and 5 "strongly agree".

With the assistance of the village chiefs and community volunteers, the questionnaires were distributed to 568 residents, using a cluster convenience sampling technique. Considering participants' comprehension ability, the survey was conducted using self-report and interview. In total, 568 questionnaires were collected and 473 questionnaires are considered valid. The valid return rate was 83.3%.

2.3. Statistical analysis

After collecting the data, SPSS version 22.0 was used to carry out the statistical analysis. Descriptive statistics and item analysis were used to check the consistency among items and among samples, as well as stability of factor structure. Then, EFA (exploratory factor analysis) was conducted to find out the factor structure of residents' sense of happiness in the neighborhood. At the same time, reliability analysis was conducted to ensure the reliability of each factor. Finally, the multiple linear regression analysis (stepwise regression) was preformed to identify physical environmental factors that significantly predict resident's happiness.

3. Results

3.1. Social-demographic characteristics of the interviewees

A total of 473 participants were surveyed. The participants' social-demographic characteristics, including gender, age, education, marital status, home ownership and living duration, are shown in Table 1.

Table 1. Sociodemographic characteristics of the respondents

Gender	N	%	Marital status	N	%
Male	217	47.3%	single	48	10.5%
Female	242	52.7%	married	354	77.5%
			widowed	36	7.9%
Age			divorced or separated	18	3.9%
20-30	24	5.1%			
31-40	52	11.0%	House ownership		
41-50	95	20.1%	Self-owned	286	61.1%
51-64	171	36.2%	Rental	67	14.3%
65-74	79	16.7%	Borrowed	25	5.3%
75-85	47	9.9%	Dormitory	85	18.2%
above 85	5	1.1%	Other	5	1.1%
Education			Living duration (years)		
Elementary school or under	45	9.6%	Under 1	25	5.3%
junior high school	28	6.0%	1-5	88	18.6%
Senior high school	166	35.3%	6-10	76	16.1%
Undergraduate	191	40.6%	11-15	87	18.4%
Graduate	40	8.5%	Above 16	197	41.6%

3.2. The factor structure of sense of happiness

The principal component analysis of EFA was used to examine the factor structure of sense of happiness. The eigenvalues were founded of a two-factor solution: "satisfied" factor and "joyful" factor. The eigenvalues of were 4.754 and 1.094, totally explaining 73.106% of variance. The results of factor loading and explained variance are illustrated in Table 2. Then, both factors are used as the criterion variables for the stepwise regression analysis. This result is consistent with Ruut Veenhoven and Mariano Rojas's theory of happiness and its components [11].

Table 2. The results of EFA about factor structure of the happiness scale.

Questions	Factor 1 Satisfied	Factor 2 Joyful	Communalities
Life is good	0.834	0.257	0.762
I am satisfied with my life now	0.829	0.295	0.775
Work gives me sense of achievement	0.786	0.208	0.661
I am willing to take any challenge	0.775	0.232	0.654
I am well satisfied with everything in my life	0.717	0.361	0.644
Getting along with others is fun	0.202	0.861	0.782
I often feel joyful	0.281	0.853	0.807
I am very pleasant	0.397	0.786	0.762
Eigenvalue (% of variance)	4.754(59.430%)	1.094(13.676%)	
Rotation SS loadings (% of variance)	3.380(42.252%)	2.468(30.854%)	
Explained variance (%)Total	73.106%		
Cronbach's alpha	0.890	0.856	
Kaiser-Meyer-Olkin	0.900		
Bartlett	0.000		

3.3. *The relationship between planned physical environment and residents' sense of happiness*

For the "satisfied" factor, the results of regression analysis shown that "community layout", "aesthetics", "transportation service", and "social service" are significant predictors (Table 3). All the afore-mentioned factors positively correlated with the satisfied factor. The explanation power of these predictors are significant: adjusted R^2=0.384, F (7.677) =74.126, p=.000 < 0.001.

As for the "joyful" factor, the model explains that "transportation service", "green area", "recreational service", and "business service" to be significantly related with it (Table 3). All have positive power: adjusted R^2=0.407, F (3.987) =81.337, p=.000 < 0.001. The standardized coefficients (β) display the effect size of predictors of the two regression to vary between small to medium.

The relationship between happiness and "physical environment" dimension can be expressed as the following two equations:

$$\text{〔Happiness: Satisfied factor〕} = 1.531 + 0.205\ \text{social service} + 0.138\ \text{community layout} + 0.158\ \text{transportation service} + 0.114\ \text{aesthetics} \quad (1)$$

$$\text{〔Happiness: Joyful factor〕} = 1.184 + 0.181\ \text{transportation service} + 0.211\ \text{green area} + 0.188\ \text{recreational service} + 0.091\ \text{business service} \quad (2)$$

According to the equations above, the effects of physical environment on residents' sense of happiness can be observed. Generally speaking, residents' sense of happiness is simultaneously influenced by the planning and design of the physical environment of the community and the services available at the community. The effect of community services on satisfied factor and joyful factor is medium, but stronger than physical environment factor. Although green area is a primary factor that influences joyful factor, but other community services factors, such as transportation service, business service, and social service also have only low to medium effect on joyful factor. Among them, transportation service and social service have significant influence on satisfied and joyful factors. Furthermore, compared with transportation service, the effect of social service to both factors is medium.

Table 3. Predictors of the happiness factors

	B	SEB	β	t	VIF
Dependent variable: "satisfied" factor					
Constant	1.531	.140		10.951***	
social service	.205	.038	.267	5.425***	1.850
community layout	.138	.054	.147	2.547*	2.531
transportation service	.158	.044	.189	3.581***	2.118
aesthetics	.114	.041	.148	2.771**	2.183
Dependent variable: "joyful" factor					
Constant	1.184	.157		7.525***	
transportation service	.181	.041	.223	4.460***	1.968
green area	.211	.037	.256	5.683***	1.598
recreational service	.188	.045	.216	4.207***	2.084
business service	.091	.045	.091	1.997*	1.646

*= $p < 0.05$ **= $p < 0.01$ ***p=< 0.001

4. Conclusion

This study showed how various of physical environment factors within the community influence residents' sense of happiness and explored improvement of which physical environment factors can improve residents' sense of happiness. Other researchers have found that residents recognize the fact that the quality of the living environment can be divided into spatial, functional and contextual factors [12-13]. This study further discovered that some physical environmental factors, including planning and design of the physical environment, and community services influence residents' sense of happiness with varying significance. The study done by Asiyeh S. in 2017 on Iranian young women about the relationship between the living environment, well-being and lifestyle behaviors has found that, in addition to psychological environment, negative appearances in the physical community environment have negative impact on women's sense of well-being and satisfaction rate of their living environment [14]. The results of this study echoed that of Asiyeh S. This study found that green area, community layout, aesthetics, transportation service, social service, among the other physical environment factors, have significant effect on residents' sense of happiness. Among them, transportation service and social service simultaneously influence "satisfied" and "joyful" factors; community layout and aesthetics have positive correlations with "satisfied" factor; green area and business service significantly influences "joyful" factor.

Numerous studies have shown that green area in a community is correlated to residents' sense of happiness8] [15-18]. This study further found that the size of green area and accessibility of the green area, the size of the trees and the landscaping, are positively correlated with "joyful" factor. Criminologists often argue that community layout is an important element of community safety. Studies have shown that community layout is directly or indirectly correlated to residents' sense of well-being. Relevant research showed that the design of street networks, particularly the permeability of community layout is key to residents' health and well-being [18-19]. In addition, this study also found that aesthetics of architecture and outdoor space in the community is key to create residents' sense of happiness and it has positive correlation with "satisfied" factor. In other words, residents living a beautiful community tend to have higher sense of happiness. Many studies have found that in nursing homes, rural areas or poor areas, the aesthetics of exterior space in the community or interior space inside the houses are significantly correlated to residents' sense of happiness. [20-22]. Transportation service is significantly correlated to "satisfied" and "joyful" factors, the two factors that show sense of happiness. Other studies have proven that when residents live closer to the busy roads, their sense of happiness will be lower. In addition, traffic congestion will definitely lower residents' sense of happiness [23-24].

The results of this study can be used as a reference for governments to make housing policies. Professionals can also use the findings of this study as reference as they implement urban planning, architecture design and landscape design to create happy communities. Future researchers can also use the results of this study as the foundation to conduct more research on conditions related to happy and livable environment.

Acknowledgements
Special thanks to Ms. Shu-yuan Cheng, Mr. Yi-chian Huang, Mr. Jin-ting Tsai, Mr. An-guo Wang, Mr. Wen-cai Wang, Ms. Ing-lin Chen, Mr. Yao-qing Feng and Ms. Wen-juan Gu for their effort in collecting data for this study.

References
[1] Wu P H 2011 *National well-being indicatior* (Taipei: Department of Statistics, DGBAS, Executive Yuan) p 3-6
[2] Thinley J 2007 'What is gross national happiness', in Rethinking Development *Proceedings of the second international conference on gross national happiness* (Bhutan: The Centre for Bhutan Studies)
[3] OECD 2011 *How's Life? Measuring Well-being* (Landon: OECD Publishing)
[4] Kamp I V, Leidelmeijer K, Marsman G and Hollander A 2003 Urban environmental quality and human well-being Towards a conceptual framework and demarcation of concepts: a literature study *Landscape and Urban Planning* 65(1-2):5-18
[5] Brereton F, Bullock C, Clinch J P and Scott M 2011 Rural change and individual well-being: the case of Ireland and rural quality of life *European Urban and Regional Studies* 18(2):203-227
[6] Phillips D R, Siu O, Yeh A G and Cheng K 2005 The impact of dwelling conditions on older

persons' psychological well-being in Hong Kong *Social Science and Medicine* 60(12):2785-2797

[7] Betts K, Lebbrandt S, and Moon H 2010 A critical review of the literature on social and leisure activity and wellbeing in later life *Ageing and Society* 31(2010):683-712

[8] Smyth R, Mishra V and Qian X 2008 The environment and well-being in urban China *Ecological Economics* 68(2008):547-555

[9] Zuniga-Teran A A, Orr B J, Gimblett R H, Chalfoun N V, Guertin D P and Marsh S E 2017 Neighborhood design, physical activity, and wellbeing: applying the walkability model *International Journal of Environmental Research and Public Health* 14(1): 607-631

[10] Lu L 1998 The meaning, measure, and correlates of happiness among Chinese people *Humanities and Social Sciences* (8):115-137

[11] Veenhoven R and Rojas M 2013 Contentment and affect in the estimation of happiness *Social Indicators Research* 110(2): 415-431

[12] Bonaiuto M, Fornara F and Bonnes M 2006 Perceived residential environment quality in middle- and low-extension Italian cities *European Review of Applied Psychology* 56(1):23-34

[13] Tu K J and Lin L T 2008 Evaluative structure of perceived residential environment quality in high-density and mixed-useurban settings: An exploratory study on Taipei City. *Landscape and Urban Planning* 87(3):157-171

[14] Salehi A, Harris N, Sebar B and Coyne E 2017 The relationship between living environment, well-being and lifestyle behaviours in young women in Shiraz, Iran. *Health and Social Care in the Community* 25(1): 275–284

[15] Houlden V, Weich S and Jarvis S 2017 A cross-sectional analysis of green space prevalence and mental wellbeing in England *Public, Environmental and Occupational Health* 17(1): 460-468

[16] Haase D, Kabisch S, Haase A and et al. 2017 Greening cities – To be socially inclusive? About the alleged paradox of society and ecology in cities. *Habitat International* 64(2017): 41-48

[17] Huang T, Wyka K E, Ferris E B and et al. 2016 The Physical activity and redesigned community spaces (PARCS) study: Protocol of a natural experiment to investigate the impact of citywide park redesign and renovation. *BMC Public Health* 16(1): 1160-1172

[18] Burton E and Sheehan B 2010 Care-home environments and well-being: Identifying the design features that most affect older residents. *Journal of Architectural and Planning Research* 27(3):237-256

[19] Hedayati Marzbali M, Abdullah A, Maghsoodi Tilaki, M J 2016 The effectiveness of interventions in the built environment for improving health by addressing fear of crime *International Journal of Law, Crime and Justice* 45(2016):120-140

[20] Weening M and Staats H 2010 The impact of a refurbishment of two communal spaces in a care home on residents' subjective well-being *Journal of Environmental Psychology* 30(2010): 542-552

[21] McIntyre L J and Harrison I R 2017 The effects of built environment design on opportunities for wellbeing in care homes. *ArchNet International Journal of Architectural Research* 11(1):138 -156

[22] Bond L, Kearns A, Mason P et al. 2012 Exploring the relationships between housing, neighbourhoods and mental wellbeing for residents of deprived areas *Public, Environmental and Occupational Health* 12(1): 48-62

[23] Mindell J S, Anciaes P R, Dhanani A et al. 2017 Using triangulation to assess a suite of tools to measure community severance *Journal of Transport Geography* 60: 119-129

[24] Smyth R, Mishra V and Qian X 2008 The Environment and well-being in Urban China *Ecological Economics* 68(1-2):547-555

The Meaning of Terrace as Social Interaction Place in Vertical Kampung

Antony Sihombing[1] and Nurul Gumay Poetri[1]

[1]Department of Architecture, Faculty of Engineering, Universitas Indonesia, Indonesia

*antony.sihombing@ui.ac.id

Abstract. The pros and cons of vertical housing development in Jakarta make me intend to raise the issue of the vertical *kampung*. In the *kampung*, there are social spaces where people interact to create a culture of *rukun* and *gotong-royong*. It is the terrace which become a container of social interaction. The purpose of this study is to reveal the meaning of the terrace in social interaction in the vertical *kampung* and its influence when the terrace room is not presented in the vertical *kampung*. The lost meaning of *kampung* when terrace is not presented in vertical *kampung* can be known through cognitive maps. Cognitive maps are the methodologies used to gain a human perception or view of life experiences in a particular place. This study involved several families in West Jatinegara Rusunawa to find out how people looked at the terrace in their life while in Kampung Pulo. Based on the cognitive maps method that has been done by the respondents, the terrace referred to by the villagers is an open space where people can see and interact with each other. Social space in the form of terraces which is needed by the *kampung* dweller, in fact is not a terrace house which is used as a place to receive guests and as a barrier between the fence and the body of the house.

1.Introduction

The government of Indonesia has built many flats provided for the villagers whose homes are in the eviction areas. The existence of the pros and cons of vertical kampung made me interested in discussing the displacement of settlements experienced by the residents. Flats built by the government have similar units, whereas in villages people have different houses according to their needs. It can be questioned whether the meaning of the village is always attached to a horizontal village or can also be felt in a vertical village? By raising the issue of the terrace as a place of social interaction in the village horizontal, the question arises as follows: What is the meaning of the terrace in social interaction in Kampung Vertikal? This study aims to reveal the meaning of terraces for Pulo villagers as a local culture and learn more about how social interaction changes take place in new settlements. The government wants to give a decent place, but people lose the meaning of the terrace.

2.Terrace as A Pragmatic Need of Human Being

According to Wiryomartono kampung is something unstructured, unplanned, is a non-formal settlement in relation to the wider social economic system without the existence of facilities, plans and networks to the urban economy [1]. The need for social interaction according to Wiryomartono [2] is supported by the opinion of Sihombing [3] that social interaction can be established with the existence of public space. Back to the view of Wiryomartono [2] that one of the practical needs that must be met is open air area, we can conclude that public spaces that can generate social interaction is an open space. In addition, there is a theory of Gehl [4] that people and activities attract the attention of others. Gehl [4] prioritizes that the time of visual interaction between humans longer to be able to establish a social interaction. This is because visually, humans are always interested in other humans.

According to Pfeifer and Brauneck [5] to produce an open space that facilitates visual interaction is to create an open space position in the middle of vertical dwelling and surrounded by the body of the building.

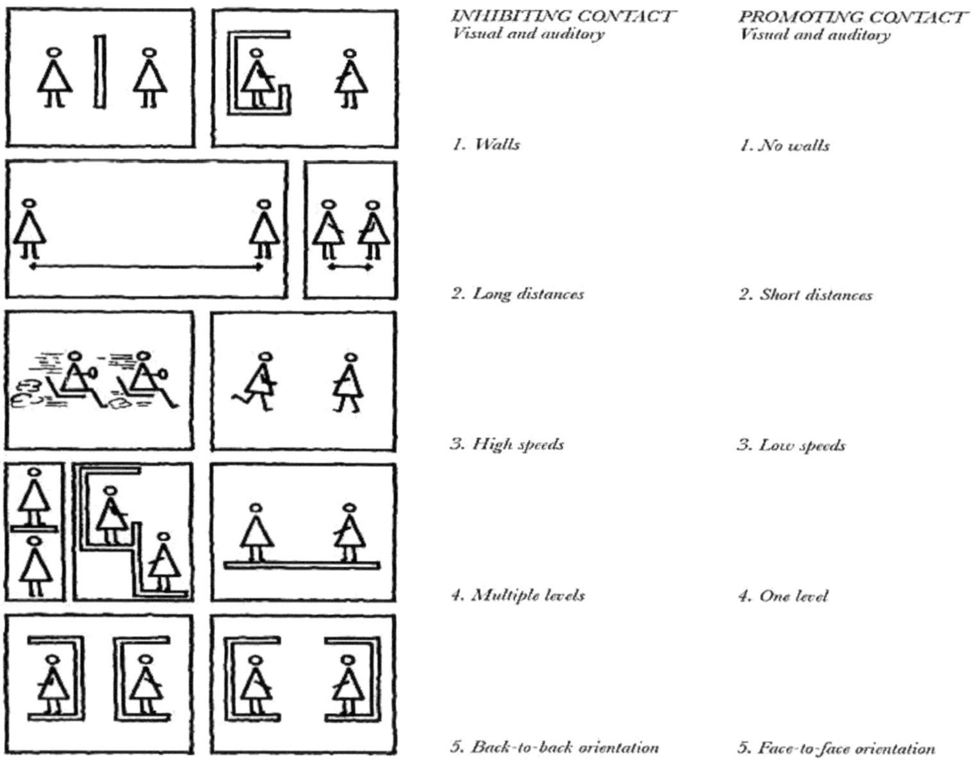

Figure 1. Visual and Auditory Contact Arrangement

Source: Gehl [4]

From the description above (Figure 1), it indicates that the good arrangement for visual contacts is if there are openings, the distance between one man to another adjacent which ideally are in one level, then also there is a factor sooner or later the activity is going on. Generally, visual interaction can occur in slow-duration activities or with longer time. Last is if the orientation between human beings facing each other (facing forward).

3.Cognitive Maps as A Research Methods

Cognitive Map is a methodology used to gain a human view or perception of life experiences in a particular place. As Hiller and Hanson said that the cognitive approach is less ambitious theoretically, of cource, because it does not aim to provide a universal theory of space; rather it is concerned to provide a methodology of investigating differences. Studies along these lines are therefore extremely valuable in providing data on differences in the ways in which individuals, and perhaps groups, cognise their environment [6]

Spatial ability becomes a spatial knowledge when movement and changes of locations in memories can be envisioned and acted upon [7]. Therefore, the method can be used in exploring the spatial memory of a family over their hometown by redrawing what happened in that space by using cognitive maps, in this study the space is a terrace in Kampung Pulo.

To achieve the objectives, this study conducted a site survey to Rusunawa Jatinegara Barat and interviewed several families who live in Rusunawa consisting of heads of households, wives, and children. Each family will describe cognitive maps which is a method to know the meaning of the terrace for Pulo villagers as a local culture and the possibility to be moved to a vertical village. The respondents was asked to draw a village map where there is a terrace in his house and drawn on a piece of paper. When describing the atmosphere of the terrace in the village, respondents will be asked to draw without copying from the real map (maps, google maps, etc.)

4.The Meaning of Terrace in Kampung Pulo

Table 1. Respondent's Identity

Number	Name	Number of Family Members	Number of Children		Family Identity		
			Woman	Man	Husband	Wife	Children
1.	Mr. Iwan Family	3	1	-	Trader	Trader	Student
2.	Mr. Sofyan Family	3	-	1	Laborer	Trader	Student
3.	Mr. Irvan Family	2	-	-	Security	Housewife	-
4.	Mr. Pendi Family	3	1	-	Driver	Housewife	Student
5.	Mr. Susanto Family	2	-	-	Employee	Housewife	-

Based on Table 1, the cognitive maps in this study were obtained from several families with different work backgrounds and ages.

IOP Conf. Series: Earth and Environmental Science **112** (2018) 012013 doi:10.1088/1755-1315/112/1/012013

Table 2. The Rekapitulation of Cognitive Maps

Number	Name	Terrace	House	Street	Plant	Door	River	Human	Stairs	*Warung*	Ditch
1.	CM 1	√	√	√	-	√	-	-	√	-	-
2.	CM 2	√	√	√	√	-	√	-	-	-	√
3.	CM 3	√	√	√	√	-	√	√	√	-	-
4.	CM 4	√	√	√	√	√	-	-	-	√	
5.	CM 5	√	√	√	√	√	-	-	-	-	√
6.	CM 6	√	√	√	√	√	-	-	-	-	-
7.	CM 7	√	√	√	√	√	√	-	-	-	-
8.	CM 8	√	√	-	√	√	√	-	-	-	-
9.	CM 9	-	√	√	-	√	-	√	-	-	-
10.	CM 10	√	√	√	-	√	-	√	-	-	-
11.	CM 11	√	√	√	√	√	-	√	-	√	-
12.	CM 12	-	√	√	-	√	-	-	-	-	-

From the results of data recapitulation based on cognitive maps in Table 2, the majority of the respondents (12/12) drew house. The data shows that a house is something important in human life. The results of cognitive maps recapitulation are also in line with the expert opinion of Kevin Lynch [8] which explains the human memory of a house. Quantitatively 12 informants narrate that their home is a safe place. They recalled that their house is a place for their warm family, both nuclear and extended family. The relationship between nuclear family and its neighbors for the villagers also very intimate. It is what makes a village with an atmosphere of kinship, affection, and mutual help.

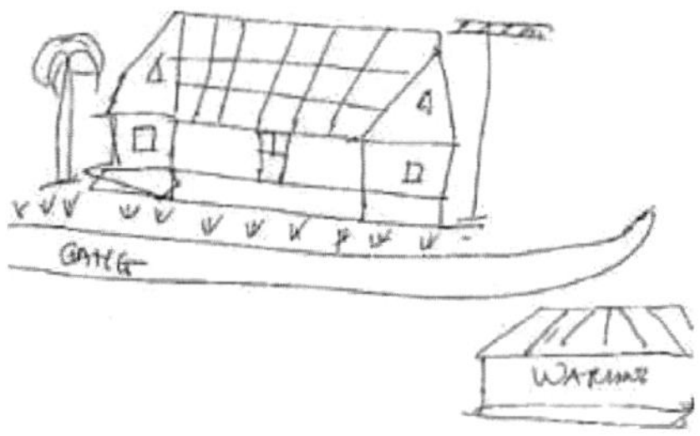

Cognitive Maps Mr. Sofyan

Cognitive Maps Mr. Iwan

Cognitive Maps Mr. Putra

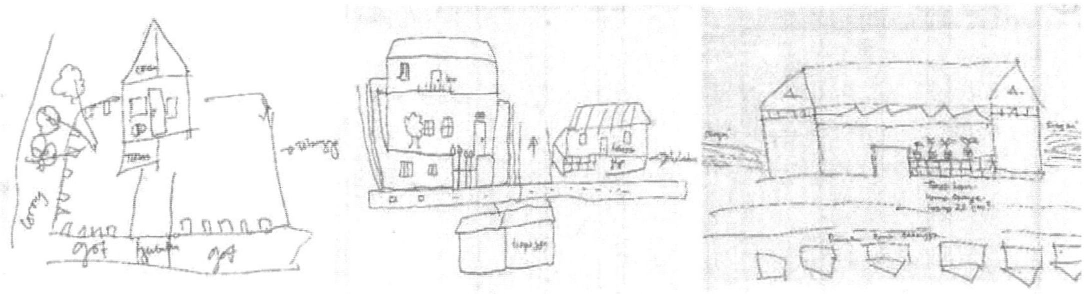

Cognitive Maps Mrs. Anisa Cognitive Maps Ms. Lina Cognitive Maps Mr. Irvan

Figure 2. The similarity of Cognitive Maps 1

These six informants in Figure 2 have similarities in their cognitive maps. They are members of the family who are often doing activities outside their home. Therefore, they often observe the state of their house from outside. The informants in this study are a laborer, trader, security officer and students.

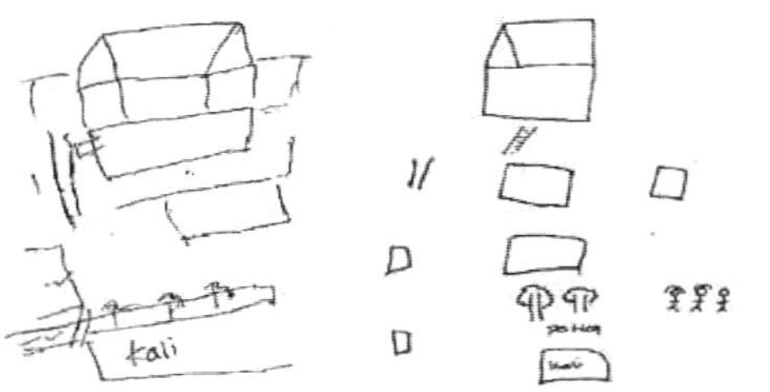

Figure 3. The similarity of Cognitive Maps 2

Figure 4. The similarity of Cognitive Maps 3

Cognitive Maps Mrs. Ida Irvan Cognitive Maps Mr. Susanto

Figure 5. The similarity of Cognitive Maps 4

Both of the informants in Figure 3 have similar cognitive maps. Both have different views about the terrace, if Ida's mother looks at the alley as a terrace then Amanda looks at the playing field as a terrace. Figure 4 describes that both maps have similarities, including having two aisles, on the right and left side of the house. The *gang* alley is the space they consider to be a terrace where they engage in social interaction. Both of these cognitive maps (Figure 5) show the difference on describe their house in their hometown. Each of them uses the overall method of the graph and the whole text. In

Indonesia context, usually the terraces in the kampung are common property. However, the definition of the terrace within the kampung itself is very wide. For example, the parking lot of the motorcycle is also considered as a terrace. Therefore, if this parking lot is removed, then there will be no interaction between families and their neighbors.

Table 3. The Meaning of Terrace for Respondent

Number	Name	The Meaning of Terrace	The Lost Meaning
1.	CM 1	Terrace as a place to chat with siblings	Family, Interaction
2.	CM 2	Alley as terrace	Friends, Interaction
3.	CM 3	Playground as Terrace	*Pengajian*, Interaction, Bravery
4.	CM 4	Terrace as a place to sit watching people pass by	Visual Interaction
5.	CM 5	Terrace as a place to greet neighbors	Neighbors, Interaction
6.	CM 6	Terrace as a place to see pets	Pets, Games
7.	CM 7	Drink coffee and chat on the river bank as a terrace	A House, Socialization
8.	CM 8	Garden on the river bank as a terrace	Land
9.	CM 9	A long bench on the edge of the alley as a terrace	Siblings
10.	CM 10	Alley as terrace	Vegetable Carts, Interaction
11.	CM 11	*Warung* in front of house as terrace	Snacks, Games
12.	CM 12	Car parking lot as terrace	Neighbors, Interaction

Based on Table 3, the cognitive maps in this study were obtained from several families with different work backgrounds and ages. I find some similarities in the maps created by each individual such as the need of space for social interaction, and the requirement of an open space where people can observe others. With various interpretation of the terrace meaning, the terrace space increases the social interaction between peoples and their surrounding. The change is felt by every member of the family who currently feels the loss of relatives, playmates, relatives and nearest neighbors, vegetable sellers, and others as I have concluded at the beginning of the discussion.

5.The Meaning of Terrace in West Jatinegara Flats

In Rusunawa Jatinegra Barat each unit has the same area of 36 m2. That information is the existence of a terrestrial meaning which is seen to be very different from a core room in general, and also the meaning of the terrace that disappeared since the source was moved to the vertical village. The terrace room that is widely known by the public is a space that is in front of the house and serves as a barrier between the body of the house and fence. The meaning of the terrace for the resource person is not a privately owned patio space right in front of the house, but they mean the terrace space becomes various places. Vertical villagers are more likely to interact in places other than those prepared by managers; they are actually doing social interactions in unexpected places (Spontaneus Space). Spontaneus Space I mean the front room of the elevator where many people meet spontaneously so as to interact like a greeting. This happens because the front room of the elevator is a meeting room of 3 activities that is waiting for the elevator, the flow of circulation of citizens from the clean market and the flow of citizen circulation from the motorcycle parking ground floor.

In the case study review of horizontal villages, it was found that the residents have the need for interacting space in the form of open space where they can see many people who are known or not. In accordance with the case study I conducted in Rusunawa Jatinegara Barat, the tower did not have the quality of the core space based on the needs of the residents while the vertical village should be designed to meet the needs of the villagers as horizontal. The various spaces mentioned by the residents as a terrace for him are the neighboring terraces in the horizontal village, the alley, the playground, the seating, the greeting place, the pet ranch, the coffee drink by the river, the riverside garden, the bench in Edge alleys, stalls and other spaces that are considered by residents as a terrace.

6.Conclusion

From the above discussion, it can be concluded that the social space in the form of terraces required by horizontal villagers is a not a porch house that is used as a place to receive guests and as a barrier between the fence and body of the house. Based on the cognitive map method that has been done by the respondents, the terrace referred to by the villagers is an open space where people can see and interact with each other. Terrace according to the residents can be various social spaces such as alleys, river banks, playgrounds, parking lots, stalls, parks and also the terrace of a neighboring house.

Kampung is an area that has a strong local culture so that when the village is moved to the vertical *kampung* then the residents need an atmosphere with the same quality of space with their local culture. *Kampung* is identical with a sense of kinship and a well-established community, based on a village reference having a social space in the form of a terrace where all citizens can engage in social interactions with family, close relatives, and neighbors.

References

[1] Sihombing, A. *Conflicting Images of Kampung & Kota in Jakarta*, LAP LAMBERT Academic Publishing, England, p. 84, 2010.

[2] Wiryomartono, B P. *Urbanitas dan Seni Bina Perkotaan*, Balai Pustaka, Jakarta, Indonesia, 2002.

[3] Sihombing, A. *Conflicting Images of Kampung & Kota in Jakarta*, LAP LAMBERT Academic Publishing, England, 2010.

[4] Gehl, Jan. *Life Between Buildings*, Van Nostrand Reinhold, New York, 1987.

[5] Pfeifer, Gunter & Brauneck, Per. *Residential Buildings*. Birkhauser Verlag, Basel, 2015.

[6] Sihombing, A. *Conflicting Images of Kampung & Kota in Jakarta*, LAP LAMBERT Academic Publishing, England, p. 216, 2010.

[7] Tuan, Y F. *Space and Place: The Perspective of Experience*, University of

Minnesota Press, Minnesota USA, 2001.

[8] Lynch, Kevin. *The Image of the City*. The Technology Press & Harvard University Press, Cambridge, Massachusetts, 1960.

Space Sharing Between Formal and Informal Sectors In Kemang Area

Antony Sihombing[1,*], Hafizka Chandra Dewanti[1]

[1] Department of Architecture, Faculty of Engineering, Universitas Indonesia, Indonesia

**Antony.sihombing@ui.ac.id*

Abstract. Sharing is a form of human activity as a social being, over resources or spaces. Humans generally define their space according to their psychological and biological needs. However, what if space sharing takes place in an urban scope? The phenomenon of space sharing happens in Kemang area done by two distinct sectors, formal and informal which both are engaged in commercial activities independently. In the study of territory and the behavior settings, the quality of environment can affect the formation of human activities in a space, occurs a phenomenon of space sharing. The researcher will discuss and present a form of the space sharing by the formal and informal sectors in a commercial area through different environmental qualities. In some circumstance, a form of space sharing can increase the value of space and obtain to a shared space, where both sectors mutually take benefits. Otherwise, it leads to an adverse impact in some others. It is clear that differences in the physical environment and social environment have an impact on the formation of space sharing and the value of space in an urban region.

1.Introduction

Kampung is one element of locality of a region, because kampong formed by its citizens for many years. However, since the early 2000s, development in the City of Jakarta increases, so that everything that considered as slum replaced with permanent buildings that are designed more structured and organize, in the hope of the realization of the ideal city. This is turns to an elite area, where the area offers medium to upper settlements, shopping centres, culinary and entertainment center day and night, there are still many street hawkers community and other middle-class social and economic activities. In fact, not infrequently these activities occur in a small scope and survive in a long time.

2. Space Sharing

Humans are social beings so that people always need others. Sharing is an act of social interaction between humans to realize harmony and to complement each other. Sharing is the shared use of resources or space. In a narrow sense refers to a merging of either alternatively limited or inherent use in human or naturally occurring activity. Henri Lefebvre (1991) in his understanding of space, expressed "social space is a social product". Space is present not only because of the social activities that fill the emptiness, but space is formed also due to social activities. [1]

Habraken argues the territory is formed from our actions in occupying space. [2]. Which explains that humans have the ability to choose the existence of other goods against themselves (1998), while Jon Lang, in his book Creating Architectural Theory, suggests that territories are formed to minimize aggression and give identity to space (1987, Chapter 14, p.148). Territory also has several functions such as security, identity, stimulation, and frame of reference. (Lang, 1987). [3]

Hall in his book entitled "The hidden dimension" (1990) expressed his understanding of the dimension space. Humans have 4 zones of distance to objects or other creatures, namely intimate distance, personal distance, and social distance. Comfortable public space has a personal zone and social distance. Allan B. Jacobs in his book The Great Streets (1995) expressed his opinion about the characteristics of physical quality of a good public space, namely; Achievement, bringing people together, publicness, liveability, safety and participation and responsibility. [4]

3. Quality of Environment Against "Space Sharing"

Roger Baker (1989) in his study on the setting behaviour explained that human behaviour is understood as a combination of human activity [5]. It is also explained by Lang that the pattern of human behaviour is the link between activity, environmental layout, and within a certain period. The activity and quality of the social environment is a key element in the process of space sharing. Jan Gehl in his book "Life Between Building" tries to express his understanding of human activities in the public sphere. He categorizes it based on the relationship of activities with basic human needs into three activities, namely: necessary activities and optional Activities and social activities. [6]

Table 1. Graphic representation of the relationship between environmental quality and the magnitude of social activity that can occur. (Source:Gehl,1980)

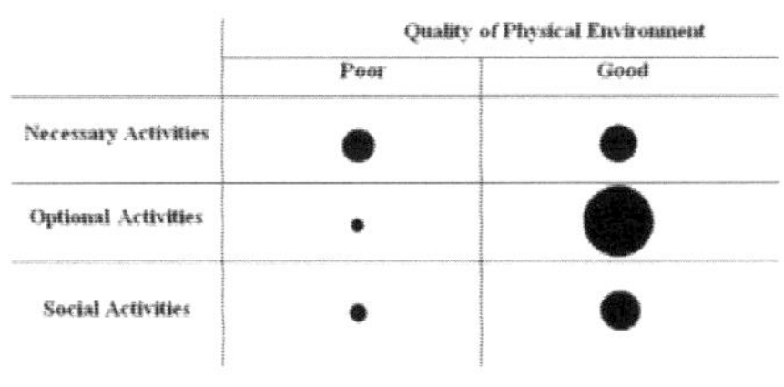

3.1 Identity of space against Space Sharing

According to Lynch region identity can be known through environmental observations. Environmental observation (environmental image) is a result of a process that occurs back and forth between the observer and the environment. In environments there are differences and interrelationships or relationships with each other, and then observers with great adaptability can use their ability to choose, organize, and make decisions about what they see. In addition the different environments will be opposed, which will then facilitate the process of building an image or the so-called identity [7]

3.2 Commercial Spaces: Formal and Informal Sectors

Outer space is an easily visible space between the formal and informal sectors, such as commercial space because it has a private level smaller than the occupancy, but larger than the public space so that commercial space is called the privatization of the public space. This space can be a place for a variety of human activities, such as a major activity place for workers or merchants, as well as a place to socialize. The activities show the process of sharing a varied space by seeing the pattern of changes in the use of the place, the social quality of which is crowded and quiet. Commercial activities have the meaning as an activity undertaken by a person either personally or entity that aims to gain profits either directly or indirectly. The dictionary of *KBBI* (*Kamus Besar Bahasa Indonesia)* explain *komersial* or commercial as everything connected with the trade, in which there is socio-economic interaction. Then proceed with the phrase "... sometimes sacrificing other values (social, cultural, etc.)". In commercial space there is a high social interaction with a high possibility to break through boundaries and values or norms. Since the essence of trading is profit gaining, hence there are sector differences especially in commercial space, which is formal sector and informal sector.

The street vendor is an example of the informal sector that is commonly found in the city of Jakarta. It defined as someone who offers goods for sale to the public without having a permanent place from what it sells. (Bhowmik 2005, in Asiedu & Mensah 2008). The street vendors have the potential to become the subject of the space sharing process, especially with a formal commercial space, because street vendors do not have a legal place in their business. In commercial space, the owner is a personal or private group who both independently conducts trading activities and generates their own benefits. Everything that is done is to make a profit, directly or indirectly, so that in commercial space sometimes breaks through the boundaries of value in order to achieve maximum profit.

4. Space Sharing In Kemang Area

4.1 Physical quality and environmental quality in Kemang Area

Kemang area located in Bangka, Mampang Prapatan, South Jakarta. The area has a hilly geographical condition and was originally occupied by locals of low density. Prior to metropolitan infrastructure and development, access to the region was limited. The area that became my observation are Kemang Selatan street, Kemang Dalam street and Kemang Raya street. Kemang Selatan Street is a mixed zone, the condition of the shops is also diverse. There is a Hero supermarket located at the end of South Kemang street which has a two-storey height and a ground floor building area of more than 500square meters, with a large parking area in front of the building to the street. This building has a boundary area using a fence, but still can be seen clearly from the road. The nearest commercial area of the informal sector is outside the fence, which is a merchant who uses a bicycle or a small wagon and is on the road to Hero, a flower trader who settles using a semi-permanent building made of wood.

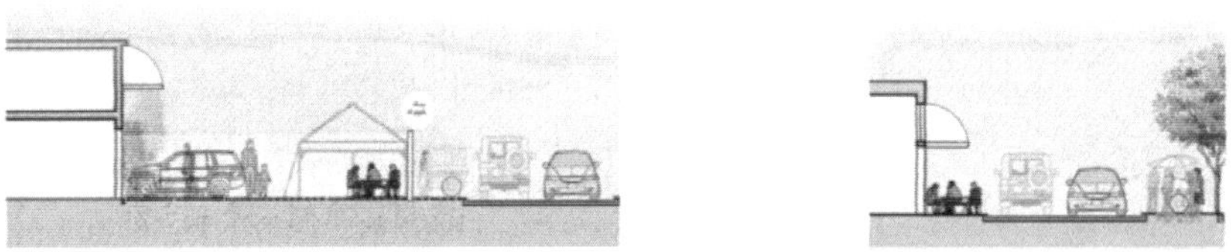

Figure 1. [Section of Kemang Selatan street (a) café 1; (b) café 2. (Source: personal processed)

Generally, the Kemang Selatan street is more crowded during the day, while at night the shops are closed and there are only a few street vendors (figure 1) . Street vendors in this street become the center of human gathering at night, because the location is fairly quiet especially around schools and shopping matters. From the picture diagram visible street vendors have a status where the road conditions are quiet, as the illustration following . (figure 2)

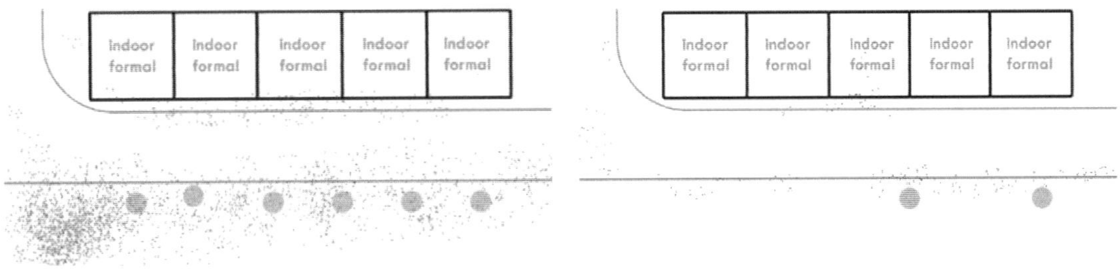

Figure 2. [section of Kemang Selatan street (a) café at 12.00 p.m. ; (b) café at 05.00 p.m. (Source: personal processed)]

In general of Kemang Raya street, the formal sector of commercial building has more than 50 square meters, with building height ranging from one to three levels. On average the store has no fence so the physical space becomes open, but there are some limiting stores with a solid fence as tall as 1.5 meters or with tall bushes. Distance between the roads with the building body from 5 to 15 meters which is used for vehicle parking area. So the existence of street vendors and small stalls can be clearly seen but the activity is not until heard or smelled. Some stores do not use fences, but they limit the territory with the security system as well as solid openings in the building. (figure 3)

Figure 3. [Section of Kemang Raya street. (Source: personal processed)]

Kemang Raya street is always crowded, especially at night until 12 pm. A lot of human activity in the outer space as well as in the buildings. The the road are always crowd, full of vehicles, and the shops always have visitors. Street vendors are seen on the sidewalks and in front of the shops.

Figure 4. [section of Kemang Dalam street (Source: personal processed)]

There are only 1-2 commercial buildings that are currently on the Kemang Dalam street because this road is a residential area. On this street the shop has a base building area of 50 square meters. The store has a border area that uses a wall as high as 2 - 2.5 meters. The entrance uses a gate with a width of 2 meters. (figure 4). The difference of the crowd between day and night at this location is not significant. It's just at night the vehicle that parked looks less than the daytime. At night it provides security to the shop, while at daytime invites the crowds of shops. (figure 5)

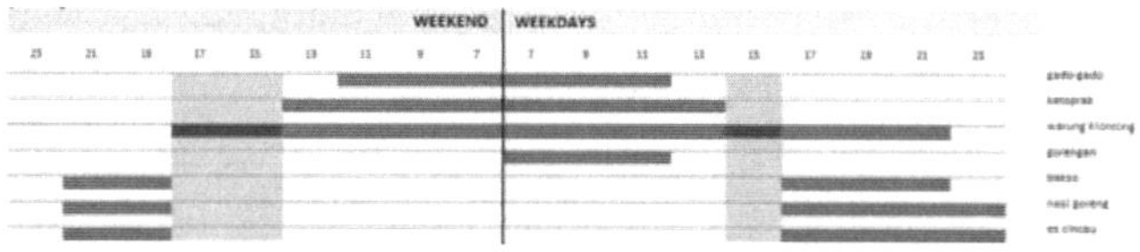

Figure 5. [Diagrams of existence of informal sectors in Kemang Dalam street (Source: personal processed on May 2017)]

4.2 Physical quality and environmental quality of the way of sharing space and its sustainability
Based on data from the study case of Kemang Raya road area, Kemang Selatan street and Kemang Dalam street, found some different kind of division of space based on divided territory.

Figure 6. [Kemang Dalam street (Source: personal processed on May 2017)]

The situation was influenced by the physical environment such as building and openness of the building to the outer space, as well as the social environment and human activities. Kemang area is a commercial area ranging from middle to lower economic level so that activities that occur in this commercial area is an optional activities and social activities. Some kinds of division of spaceinclude:

(a) (b) (c)

Figure 7. [illustration plan of The informal sector within the formal sector commercial space] (a) within the formal sector commercial space. (b) within outside of the territory bounds of formal sector in commercial space. (c) within the territory bounds of formal sector in commercial space. (Source: personal processed)]

a) The informal sector within the formal sector commercial space
Under these conditions informal sector areas are found within the formal sector area where overlapping sections become common space between the formal and informal sectors. In the literature discussed in the previous chapter by Robert S. Feldman (1989), the smaller the marked territory the more private the space is. This does not apply in this case, where the store is a space that has a large enough territorial boundary, but in fact it is the social environmental factors and human actions that occupy the space which then makes this space crowded and reduce the intensity of private owned by the physical building, And build a blurred-territory, where the territory between the formal and informal sectors becomes vague. In accordance with the Habraken phrase, territory can be formed

from our actions in occupying space. This explains that humans have a stronger ability to choose the existence of other goods against themselves.

b) Informal sector within outside of the territory bounds of formal sector in commercial space

The condition of this division of space is found in the area of Kemang Selatan and Kemang Raya. Where the physical condition of the building is very large and restricted territory or otherwise, the store area is too open, but the store has a small size and large openings.

c) Informal sector within the territory bounds of formal sector in commercial space

This category is commonly found when formal sector commercial areas have solid territorial boundaries such as Kemang Dalam street. In a region where the majority of these elite dwellings are the existence of street vendors in the streets inviting crowds on the street. So that the existence of the store becomes easily legible among the functions of the region as elite dwellings tend to be quiet. As for street vendors and *klontong* stalls, store territory used as a place to sell. This phenomenon is almost identical to Category A, which is equally beneficial, but in the space division the informal sector does not intervene too much into formal sector territory because formal sector stores

From the category of space sharing that I have described previously one by one, there is a connecting line between the three categories, namely the physical and environmental quality, such as the data table of the following observations.

Table 2. Space sharing against the quality of environment between formal and informal sectors in Kemang Area. *(Source: Personal Processed)*

	quality of environment \ location	kemang raya	kemang selatan		kemang dalam
physical	openess of publicity	open	open	close	close
	area allocation	commercial area	mixuse area	mix-use area	residential
	type of streets	major road	two way street (small)		two way street (median)
	pedestrian way	v	-		-
social	morning	oooo	oooo	o	oo
	noon	oooo	ooooo	oo	oooo
	afternoon	ooooo	oo	o	o
	evening	ooooo	ooo	ooo	ooo
space sharing category		A	B	A	C
sustainability		Temporary	constantly	constantly	constantly

If linked between the previous analyses, the physical quality of the building is related to the social environment. If the physical quality is not good, in the sense of lack of openness of commercial land, and allotment of land around is not oriented to crowd, such as residential and others, the quality of social environment becomes less, because the location of commercial which is not strategic and the visitor is few. In such conditions the informal sector shares the space at the territorial boundary to enter the territory of the formal sector space. And it happens continuously. If in a condition where the social environment and physical quality are good, then the social activity that occurs is greater, according to Jan Gehl (2011) diagram of human activity in outer space. So the formal sector does not need help on improving the social quality of the informal sector and the sharing of space does not last for long.

5. Conclusion

Based on case studies and analysis, the existence of space sharing between formal and informal sector can be influenced by the social and physical environment. The elements of environment such as building openings, building area, building area and the quality of the surrounding environment and social environment such as the utilization of space and buildings around, roads, and so forth. In large stores such as those on Jalan Kemang Raya, which has an open parking lot, informal traders can enter the territory of the shop. While on Kemang Selatan I street with 4x4 meter shop having big openings and dining place on the terrace that directly intersect with the street, informal traders can be present away from the boundaries of the shop territory like on the street (outside the store physical territory). The higher the social quality that occurs independently in a commercial location then the time for the continuity of sharing space by the formal and informal sectors becomes less, incontinent or temporary. While in a condition of poor social environment independently, then the continuity of sharing space becomes longer, because sharing space increases social interaction. Different criteria of space elements can also cause different ways of sharing space within the commercial sphere. In the case of Kemang area and the analyzes obtained and the resulting benefits, I divide them into several types based on the benefits received by each sector, namely:

5.1. Not mutually benefical

In this case the formal and informal sectors do not share space in their territories. However, mutual expansion of territory by visitors, which extends to the territory does not occur in the territory, but in public space, such as roads.

5.2. Profitable one

This phenomenon occurs at kemang raya where informal sector traders take advantage of formal sector storefront areas, whereas shops do not benefit from the presence of informal sector traders. In this case it is often a conflict where the formal sector is disturbed by the presence of informal traders in its territory, so users sometimes take action to defend their territory. The existence of the informal sector in the shared-space persists in a short time and tends to change, move, and have no pattern.

5.3. Simbiosis Mutualism Shared-Space

In sharing common space, between the formal and informal sectors both provide benefits (symbiotic mutualism). The existence of the informal sector generally benefits the store because it has the power to invite crowds and human activities in the outer space, but traders do not have any areas of power. While the formal sector has an area of power but can not independently invite activities and visitors so that they complement each other. Based on observations and interviews, this category of space sharing can last for quite a long time.

References

[1] Lefebvre, Henri. 1991. *The Production of Space*. Trans. D. Nicholson-Smith.
[2] Hebraken, N J. 1998. *The Stucture of the ordinary*: form and control in the built environment. Jonathan Teicher(ed). USA: The MIT Press
[3] Lang, Jon. 1987. *Creating architectural theory: The role of the behavioural sciences in environmental design.* New York: Van Nostrand Reinhold.
[4] Jacobs, Allan B. 1993. *The Great Streets*. Cambridge: The MIT Press.
[5] Estherina, Ika. 2009. *Irama di suatu kawasan.* Depok: Perpustakaan UI.
[6] Gehl, Jan. (1987). *Life Between Buildings*, Van Nostrand Reinhold, New York.
[7] Lynch, Kevin. (1960). *The Image of the City*. The Technology Press & Harvard University Press, Cambridge, Massachusetts.